Energy and Entropy

Michael E. Starzak

Energy and Entropy

Equilibrium to Stationary States

 Springer

Michael E. Starzak
Department of Chemistry
State University of New York
Binghamton
Binghamton NY 13902-6016
USA
mstarzak@binghamton.edu

ISBN 978-1-4899-8367-1   ISBN 978-0-387-77823-5 (eBook)
DOI 10.1007/978-0-387-77823-5
Springer New York Dordrecht Heidelberg London

Printed on acid-free paper

Springer is part of Springer Science+Business Media (www.springer.com)

Preface

The study of thermodynamics is often limited to classical thermodynamics where minimal laws and concepts lead to a wealth of equations and applications. The resultant equations best describe systems at equilibrium with no temporal or spatial parameters. The equations do, however, often provide accurate descriptions for systems close to equilibrium. . Statistical thermodynamics produces the same equilibrium information starting with the microscopic properties of the atoms or molecules in the system that correlates with the results from macroscopic classical thermodynamics. Because both these disciplines develop a wealth of information from a few starting postulates, e.g., the laws of thermodyamics, they are often introduced as independent disciplines. However, the concepts and techniques developed for these disciplines are extremely useful in many other disciplines. This book is intended to provide an introduction to these disciplines while revealing the connections between them.

Chemical kinetics uses the statistics and probabilities developed for statistical thermodynamics to explain the evolution of a system to equilibrium. Irreversible thermodynamics, which is developed from the equations of classical thermodynamics, centers on distance-dependent forces, and time-dependent fluxes. The force flux equations of irreversible thermodynamics lead are generated from the intensive and extensive variables of classical thermodynamics. These force flux equations lead, in turn, to transport equations such as Fick's first law of diffusion and the Nernst Planck equation for electrochemical transport.

The book illustrates the concepts using some simple examples. These examples provide a physical basis that facilitates understanding of the more complicated systems. Probabilities and averages in statistical thermodynamics are developed for systems with only two or three energy levels. The effects of interactions can be demonstrated effectively with such systems. The same techniques are then applied to continuous systems such as Maxwell Boltzmann velocity and energy distributions. The probability that a molecule has sufficient energy to react is developed using the same techniques.

Some models are developed within different disciplines (chapters) to contrast the different approaches. The Ehrenfest urn (or dog-flea) model, originally used to contrast states and distributions in statistical thermodynamics, is also used to describe the kinetic approach to equilibrium and the stationary state produced by a directional flow of particles through the system. Bose Einstein statistics are introduced in statistical thermodynamics and then applied in kinetics to calculate the probability that sufficient energy will collect in a bond to rupture it.

The first law of thermodynamics in Chapters 1–3 is expressed as a generic equation that leads easily to specific applications such as adiabatic and isothermal expansions of both ideal and real gases. Entropy is considered from both thermodynamic (Chapter 4) and statistical (Chapter 5) points of view. The statistical chapter includes some information theory. It also presents some paradoxes, which, when explained, help clarify the statistical nature of entropy.

The free energies and their physical significance are presented in Chapter 6 and applied in Chapter 7 for Maxwell's equations and thermodynamic equations of state. These equations are not limited to a study of gases. They are used to develop equations such as those for electrocapillarity and adiabatic demagnetization. The thermodynamics of solutions is developed through partial molar quantities and chemical potential in Chapter 8. Balance of the chemical potential at equilibrium is used to develop solution equations for freezing point depression and osmotic pressure and other equations such as the barometric and Clausius Clapeyron equations. The chemical potential balance equations for ionic systems also lead to equations for the electrochemical potential and the Donnan equilibrium.

Statistical thermodynamics begins with some mathematical background. The most probable distribution is illustrated for systems with two or three particles. The most probable distributions are then developed for some simple systems before the introduction of the method or undetermined multipliers that leads to the Boltzmann factor for molecules with different energies. Bose Einstein and Fermi-Dirac statistics are introduced and contrasted with Boltzmann statistics using some minimal state models.

The use of the Boltzmann factor for energy is illustrated with systems having two or three energy levels to illustrate probabilities and averaging techniques in Chapter 11. The Boltzmann factor in chemical potential is illustrated with enzymes having one or two binding sites. The probabilities and average values for these enzyme systems are compared with the same parameters developed from equilibrium macroscopic binding equilibria.

Probabilities and averaging techniques developed for discrete systems are expanded and used for continuous energy systems. The connection between a sum over Boltzmann factors for two or three energy states and the integrals over Boltzmann factors for a continuum of energy states is emphasized. The equation developed to determine average energy from the partition function for two or three states is shown to give the average energy for the continuum of states once the partition function is known. Phase space is introduced as well as the connection between quantum and classical two-dimensional systems. .The average energy for a dipole in an electric field is determined from the relevant partition function to generate the Langevin equation for the average dipole moment.

Basic interactions are illustrated in Chapter 15 using enzyme/substrate systems. Transfer matrices are introduced to describe larger one-dimensional systems with nearest neighbor operations, and the continuum Debye Huckel theory is introduced for interactions between a charged surface and ions in solution. The one-dimensional analyses serve as a precursor to more complicated two and three-dimensional systems.

Chemical kinetics merges thermodynamics with time. The Ehrenfest urn model for systems at equilibrium is expanded to show the temporal evolution to equilibrium. The statistics of this temporal evolution are developed using stochastic methods. Bose Einstein statistics are used to establish the number of states in unimolecular reaction theory and the probability that sufficient energy will accumulate in a single bond for reaction. Bose–Einstein statistics are also used to develop probabilities for energy transfer between molecules.

The irreversible thermodynamics in Chapter 16 develops a generalized force flux equation and uses it to develop some illustrative pairings. The velocity of an irreversibly expanding piston, for example, is equivalent to a flux as defined by the generalized equations. Force–flux equations are used to develop transport equations such as Fick's diffusion equation and the Nernst–Planck electrodiffusion equation. The Goldman equation, the constant field solution of the Nernst–Planck equation is developed for both ion and charge fluxes, and these two equations are used as a specific example of the Onsager reciprocal relations. The coupling of vector (diffusion) and scalar (chemical reaction) fluxes is illustrated using the transport of ions through membranes as aion/ionophore complex.

The final chapter includes some simple flow models that open avenues to some of the more active areas of modern thermodynamics, transport, and chemical research. Such models illustrate the overall reduction in entropy in a system held at steady state by a flow. Simple oscillatory systems such as nerve action potentials and oscillating reactions are described.

The book is intended to provide a solid grounding in both the mathematical techniques and the physical concepts they describe. Chapter problems are directed toward developing and using new equations to show that the reader has mastered the techniques developed in the chapters. The text lays the framework to help the student proceed to more advanced treatises on the material.

This book was developed and refined from my courses on thermodynamics, statistical thermodynamics, and kinetics. I am indebted to the students whose perceptive questions and insights helped me find the most effective ways to present the material.

Contents

Chapter 1
The First Law of Thermodynamics

1.1 The Conservation of Energy

The first law of thermodynamics is the law of conservation of energy; energy is neither created nor destroyed. It can occur in different forms such as thermal or electrical energy and can be converted between these forms. Since matter and energy are intra-convertible, matter itself might be considered an energy. For most situations, however, the energy converted from matter is negligible and is ignored.

Energy is also transferred between locations. The preferred approach in thermodynamics is to select a specific location called the system and record energy in all its forms within that system. Everything outside the system boundaries is the surroundings. Energy transfers between the system and surroundings conserve energy. A 200 J energy loss from the system equals a 200 J energy gain for the surroundings. The energy change for the universe, the sum of energies for the system and surroundings, is always zero.

A sealed glass bulb with gas is a system. The gas can lose or gain energy. The glass bulb is assigned to either the system or surroundings if its energy does change during any transfers. A balloon or a biological cell is surrounded by a flexible surface. Expansion of this surface requires energy and the surface can be included as part of the system.

Since the surroundings include everything but the system, its energy total should be difficult to monitor. However, the surroundings can be approximated by a region adjoining the system. A glass bulb is immersed in a water bath at some temperature. An energy loss from the system causes an equal energy gain entirely in the water bath to a good first approximation if the bath is large enough.

Energy and energy changes occur in different forms that can be monitored by system variables. The temperature, pressure, and volume of a gas must be known so that any observer can prepare the system in exactly the same way. Although the surroundings can also be characterized by variables such as temperature, special emphasis is placed on these state variables within the selected system. These state variables are often related through equations of state. The ideal gas law for pressure P, volume V, temperature T, and moles n

$$PV = nRT$$

M.E. Starzak, *Energy and Entropy*, DOI 10.1007/978-0-387-77823-5_1, © Springer Science+Business Media, LLC 2010 connects these four state variables in a fairly accurate description of gas behavior. The direct proportion between energy and temperature for an ideal gas is as:

$$E = CT$$

where C is constant is another equation of state.

The transfer of energy is used to calculate the total energy change in the system. This change in energy is more useful and more easily determined than the absolute energy of the system. The change in system energy can be determined in two distinct ways: (1) monitoring the loss or gain of energy of the system and (2) observing changes in the other system variables that correlate directly with the energy change of the system.

Heat is a non-directed transfer of energy to or from the system. A bulb of gas at 400 K dipped into a water bath at 300 K will transfer energy equally across all the bulb surfaces. The energy transfer is heat. In this case, heat is lost from the hot system to the surroundings as the system cools. The system acts as the reference so that heat loss from the system is negative.

When a hot ball is dropped into a water bath, the internal energy decreases by ΔE because an amount of heat q is transferred from the system to the surroundings. Internal energy is used for the change within the system, while q is the energy transferred randomly across the system boundary. Heat and the change in internal energy are equal:

$$\Delta E = q$$

"delta" defines macroscopic changes within the system. An infinitesimal change in internal energy, dE, is equal to an infinitesimal transfer of heat $dE = dq$.

Directed transfers of energy to or from the system are defined as work with units of energy. A change in the internal energy is the sum of both heat and works transferred:

$$\Delta E = q + \sum w_i$$

Work can include pressure volume work of expansion, electrical work, etc.

The differential equation for heat transfer distinguishes two types of differential:

$$dE = \delta q$$

dE is a perfect differential. A definite integral between limits E_1 and E_2

$$\int_{E_1}^{E_2} dE = E_2 - E_1$$

depends only on the initial and final limits and not the path taken between limits. δq, by contrast, depends on the path.

Heat can be added to a system by placing it in a water bath (the surroundings) at a higher temperature. The heat added can depend on the temperature difference. However, no matter how the heat is added, the internal energy for an ideal gas depends only on the temperature change of the system. The temperature of the system is easily measured. Once the temperature change and, for non-ideal gases, changes in other system variables are known, the internal energy is determined independently of the manner in which heat and work are added.

1.2 Molar Heat Capacities

Different materials can absorb different amounts of energy. The specific heat of a material is the energy absorbed by 1 g of this material when the temperature increases by 1°C. The molar heat capacity of this material is the energy absorbed by 1 mol when the temperature increases by 1°C. A specific heat or molar heat capacity is nearly constant over a limited range of temperatures.

The empirical law of Dulong and Petit states that atomic solids have a molar heat capacity of 25 J deg^{-1} mol^{-1} ($3R$ with $R=$ 8.31 J mol^{-1} K^{-1}). Gold atoms with a higher density and lower specific heat have the same molar heat capacity as aluminum atoms with lower density and higher specific heat.

The empirical law of Dulong and Petit is used to estimate atomic molar masses. The specific heat of the metal is determined experimentally and divided into the molar heat capacity to give the atomic molar mass. An atomic metal with a specific heat of 0.415 J °C^{-1}g has an atomic weight:

$$\text{amm} = \frac{25\,\text{J}\,\text{C}^{-1}\,\text{mol}^{-1}}{0.415\,\text{J}\,\text{C}^{-1}\text{g}^{-1}} = 60\,\text{g}\,\text{mol}^{-1}$$

Monatomic gases such as argon or helium, despite their mass difference, have approximately equal, constant molar heat capacities of 12.5 J deg^{-1} mol^{-1} ($3R/2$) at constant volume. Diatomic gases, such as N_2 or O_2, have molar heat capacities of about 20.8 J deg^{-1} mol^{-1} ($5R/2$) at room temperature and constant volume.

The pattern in heat capacity values is related to locations for storing energy within an atom or a molecule. Atomic motion in a three-dimensional space is described by translational kinetic energies in three independent directions x, y, and z:

$$\varepsilon_x = \frac{1}{2}mv_x^2$$

$$\varepsilon_y = \frac{1}{2}mv_y^2$$

$$\varepsilon_z = \frac{1}{2}mv_z^2$$

The atomic gas heat capacity of $3R/2$ is generated if each of these three independent energies contributes a heat capacity $R/2$ to the total. Since the product of the heat capacity and the temperature has units of energy, e.g., J mol^{-1}, each independent motion or degree of freedom stores an equal internal energy

$$E_i = \frac{R}{2}T$$

This is the equipartition of energy theorem. The internal energy change for 1 mol of a monatomic gas with three independent motions (x, y, z) is

$$\Delta E = 3RT/2$$

If the atom is confined to a two-dimensional surface, e.g., a table, its heat capacity would be only $2R/2$. Energy is stored in only two translations.

Diatomic molecules, such as O_2 or N_2, have larger heat capacities because, in addition to the three independent translations, these molecules can rotate about only two of three perpendicular axes; energy is not easily transferred to rotation about the internuclear axis for a linear molecule. Each independent rotation (degree of freedom) has an energy:

$$\varepsilon = \frac{1}{2}I\omega^2$$

The five degrees of freedom (three translations plus two rotations) each contribute $R/2$ to the total heat capacity so that $C_v = 5R/2$. A linear molecule such as CO_2 also has five degrees of freedom (no rotation about the long axis) and a heat capacity of $5R/2$.

A non-linear molecule can rotate about three perpendicular spatial axes so that it has three translational and three rotational degrees of freedom at room temperature. If vibrational degrees of freedom do not absorb energy at this temperature, the total molar heat capacity for such molecules is

$$C_{tot} = C_{trans} + C_{rot} = 3\frac{R}{2} + 3\frac{R}{2} = 3R$$

The atoms in a solid are confined to a specific location in the crystal and so can neither translate nor rotate. They are, however, free to vibrate about a fixed center at this site in three distinct directions. The energy for one of these vibrations is the sum of a kinetic energy and a potential energy,

$$E = 1/2mv^2 + 1/2kx^2$$

where m is the mass of the atom moving at velocity v and x is the displacement of the atom. The force constant k is a measure of the restoring force experienced by the vibrating molecule.

For vibrations, each squared term in the energy expression defines a degree of freedom for an atom in the crystal. Each vibration for a mole of atoms has $C = 2(1/2R)$. The three possible spatial vibrations have six degrees of freedom and $6(R/2) = 3R$, the empirical heat capacity for atomic crystals in the law of Dulong and Petit $(3 \times 8.31 \text{ J mol}^{-1} \text{ K}^{-1} \approx 25 \text{ JK}^{-1} \text{ mol}^{-1})$.

The energy is determined from these heat capacities:

$$E = CT$$

An atom of larger mass might be expected to store more energy. However, since $E = \frac{1}{2} Mv^2$ for a mole of atoms, the velocities of atoms of different mass must differ to maintain constant energy at constant temperature. Since

$$E = 3/2RT = \frac{1}{2}Mv^2$$
$$V = \sqrt{3RT/M}$$

The atoms of larger mass have a lower velocity to give the same energy. Energy, not mass, is the primary variable.

Although the equipartition of energy theorem predicts a constant $C = R/2$ for each degree of freedom, this heat capacity can be temperature dependent for some systems. For example, (1) the heat capacity of an atomic crystal decreases at lower temperatures and approaches zero as the temperature approaches $0K$ and (2) the heat capacities of diatomic molecules change from $5R/2$ to $7R/2$ as temperature increases. These are quantum phenomena. For solids, energy is added to the vibrations in discrete packets (quanta). At lower temperatures, there are fewer packets with sufficient energy to produce the discrete change in vibrational energy so that less energy is absorbed by the crystal and the heat capacity at lower temperatures is lower.

The nitrogen molecule is "stiff." In classical thermodynamics, the vibration absorbs with $C = 2R/2$ just like a molecule that vibrates easily. However, as a quantum molecule, a large discrete quantum is required for vibration. At lower temperatures, including room temperature, such packets are rare and the molecule remains vibrationally unexcited. With increasing temperature, quantum packets of sufficient energy are more prevalent and vibrational modes can absorb energy to increase the heat capacity by two vibrational degrees of freedom $(2R/2)$ to $7R/2$.

1.3 State Variables and Equations of State

Heat transfer into a system changes E but can also produce changes in other system variables like the temperature and the volume of a gas. Temperature, volume, pressure, and the internal energy of a gaseous system are all examples of state variables that characterize the system.

The state variables for a system can vary. The earth's gravitational field acts on the molecules of a system to produce a potential energy. However, if the system remains at the same height, this gravitational energy is constant and height h, a possible state variable, need not be included in the analysis.

A change in one-state variable can cause a change in a second-state variable. The relationship between two or more state variables is an equation of state. The ideal gas law,

$$PV = nRT,$$

relates the pressure P, temperature T, volume V, and the number of moles, n, within a system. If three of these four variables (the independent variables) are known, the fourth, dependent variable is determined from the others using the equation of state. The calculated value of the dependent variable depends on how accurately the equation of state describes the behavior of the system.

The internal energy can also be related to system variables like temperature, volume, and moles for a given system. For an ideal gas, the internal energy depends only on temperature and moles and the equation of state $E = nCT$ suggests that gas energy is independent of other state variables like volume.

For real gases, the energy can depend on temperature, volume, and moles and a new equation of state is necessary. The pressure need not be included as an additional independent variable if it can be determined from an equation of state for the gas.

Each independent equation of state that relates the state variables of the system reduces the number of independent variables by one. Holding a state variable X constant during an experiment gives $dX = 0$.

The ideal gas law approximates the behavior of the actual gas. The law predicts a volume when pressure, Kelvin temperature, and moles are known. The volume calculated might differ slightly from the actual volume and the equation of state must be refined for a more accurate fit.

The equations of state for gases have the thermodynamic variables P, T, and V. Other systems might have different state variables and equations of state. For example, the tension (or force) τ generated by a polymer, e.g., rubber band, when stretched is proportional to the temperature and the distance L the band is stretched from its equilibrium, unstretched length L_0

$$\tau = aT (L - L_0)$$

with constant a. This equation parallels Hooke's law (another equation of state) for a spring displacement, x,

$$F = k\Delta x$$

where k is constant. The equation of state for the rubber band connects T, L, and τ.

The Curie–Weiss law relates the amount of magnetization M induced in a sample to the size of the externally applied magnetic field H and the temperature T as

$$M = \frac{CH}{T - T_c}$$

where C and T_c are constants in this equation of state.

1.4 Non-Ideal Gases

The ideal gas equation

$$PV = nRT$$

is actually a limiting equation where the gas diameter approaches zero and interparticle interactions are absent. The equation describes gases at lower densities where their size is small relative to the total volume and they are separated sufficiently to minimize interactions.

A mole of real gas occupies a finite volume, approximately the volume the gas would occupy as a condensed liquid. This finite "liquid" volume (b) is incorporated into the van der Waals equation of state. The ideal gas law is predicated on a particle of zero volume free to move everywhere in the state volume V. The van der Waals equation of state excludes the volume nb of n moles of the molecules. The actual volume V is reduced to a theoretical volume available to the molecules, $V-nb$.

In an ideal gas, the pressure is produced by elastic collisions with the container walls. The change in the direction of momentum on collision produces the force and pressure. The particle transfers no energy to the wall during these elastic collisions. Collisions between particles are also elastic, i.e., no energy is lost to the internal motions of the molecule. The total kinetic energy is preserved. A non-elastic collision results for a particle that flattens or distorts when it strikes the wall. Some of the kinetic energy is expended to distort the particle.

Molecular interactions reduce the force with which a molecule strikes a wall to reduce the total pressure on the wall. A particle with a clear path to the wall is held back by interactions with particles behind it to reduce its striking force and pressure. The observed pressure P on the wall is corrected by adding back the pressure lost to particle interactions

$$P + n^2 a / V^2$$

where a is a constant related to the magnitude of the interactions for the particles. The additive pressure correction is proportional to the square of the molar density n/V of the particles since each particle interacts with the other $N-1$ particles in the system. $N-1 = N$ for large numbers of particles. The N particles then have $N(N-1) = N^2$ interactions. This translates to a term in the square of the density, $(n/V)^2$.

Excluded volume b and the interaction constant a are incorporated into the van der Waals equation of state

$$\left(P_{exp} + a\frac{n^2}{V^2}\right)\left(V_{exp} - nb\right) = nRT_{exp}$$

where the subscript "exp" is added to emphasize that these variables are the laboratory variables.

The van der Waals equation is a third order in V and a $P-V$ plot (a hyperbola for an ideal gas) has a maximum and a minimum in the intermediate volume region (Fig. 1.1) because the equation describes a continuous function with no sharp changes in slope

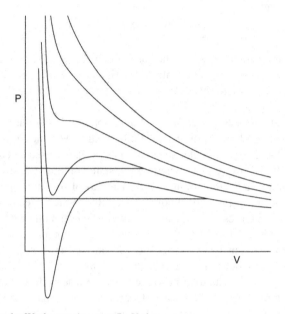

Fig. 1.1   The van der Waals equation on a $P-V$ plot

A $P-V$ plot for a real gas has three regions. The high volume low pressure region is described by the ideal gas law since $nb<<V$ and $a/V^2 << P$ in this region. As the volume approaches b, the volume is that of a liquid and the plot rises steeply. The intermediate region that describes the phase transition between the gas and the liquid is a horizontal line for a real gas. Vapor condenses to liquid, while the pressure remains constant. The continuous van der Waals equation cannot reproduce this horizontal line. However, since the equation does describe the gas and liquid portions of the $P-V$ plot, a horizontal line (the Maxwell construction) replaces V^3 dependence predicted by the van der Waals equation in this region (Fig. 1.1).

One mole of a van der Waals gas has an equation of state

$$P = \frac{RT}{V-b} - \frac{a}{V^2}$$

The Berthelot equation with a new constant a

$$P = \frac{RT}{V - b} - \frac{a}{V^2 T}$$

includes an inverse temperature dependence in the correction term.

A virial expansion is a sum of correction factors in increasing powers of some state variable. The ratio

$$\frac{PV}{nRT} = 1$$

describes the ideal gas. Deviations from ideality increase as the density of the gas increases because the effects of molecular volume and molecular interactions increase as the gas particles are closer together on average. The deviations are expressed as a sum of addition terms in powers of the density:

$$\frac{PV}{nRT} = 1 + B\frac{n}{V} + C\left(\frac{n}{V}\right)^2 + \cdots$$

The empirical constants, B, C, etc., depend on temperature but are independent of the density of the gas. The terms in higher powers of density are more accurate refinements to describe the gas behavior. The first virial coefficient, $B = B(T)$, the largest, is negative at lower temperatures and becomes positive as the temperature increases. B must then be zero at one temperature. If the gas is studied at this particular temperature, it will behave like an ideal gas to a very good approximation since the first virial term is zero.

All these non-ideal gas equations are very similar to the ideal gas equation because the correction terms or parameters are generally very small. The correction terms fine tune the ideal gas equation to describe small variations between different real gases.

1.5 Work

An automobile engine does work when gasoline in an engine cylinder is ignited and energetic product gases move a piston in the cylinder. The piston is connected to a crankshaft to convert this linear motion into rotary motion. This mechanical energy is not lost as the wheel turns; it is conserved in other forms. For example, the bulk of the work produced by the moving piston (50%) pushes air away from the front of the moving car. About 33% of it appears as heat in the engine block (the surroundings for the piston). The remainder reappears as frictional heating, either in the engine parts or at the interface between the tires and the road.

Thomson first noted the direct conversion of work energy to heat using the heat produced from a known quantity of work. Work is not lost; it is degraded to heat. Total energy is conserved.

Energy transfer as work is illustrated for a gas in a cylinder (the system) with a movable, frictionless piston (Fig. 1.2).

An applied force and pressure on the piston is produced by weight of masses on the upright piston or the weight of a column of air above the piston. For mass m, the

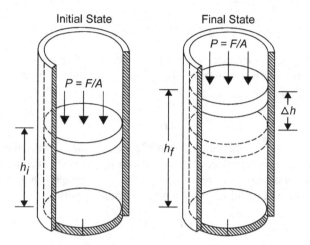

Fig. 1.2 A cylinder for pressure volume work

force mg where g is the acceleration of gravity divided by the area A of the piston is the pressure mg/A. Rather than calculate the mass of a column of air, the pressure of this column of air is operationally defined as 1 atm. In either case, work is done by the system when the volume of the gas expands to raise the piston. Work done by the system, i.e., an expansion of the gas in the cylinder, is negative; energy is transferred from the gas in the cylinder to the exterior.

Work is force × the distance Δh the piston moves

$$W = mg\Delta h$$

Even though this work for increasing x is done by the molecules in the cylinder, work is defined for energy transferred from the system to the surroundings on expansion. This work against external weights or external atmospheric pressure is the transferred variable

$$w = -F_{\text{ext}}\Delta h$$

The minus sign shows that energy leaves the system when h increases.

If h is negative (a compression), work is positive. The energy of this compression reappears in the gas of the system to increase its internal energy.

The transfer of heat q is an independent process. Heat or work is positive for a transfer into the system and negative for a transfer from the system. With this convention, a change in internal energy results from transfer of either heat or work,

$$\Delta E = q + w$$

The primary system variable is the net change of internal energy.

The external weight divided by the area of the piston defines an external pressure

$$P_{ext} = F_{ext}/A$$

The work expression can be converted into an external pressure–volume expression by multiplying and dividing by the piston area A. The product of A and the distance moved h is the volume change of the gas within the system. A differential change in h produces a differential change Adh in volume and a differential work

$$dw = -F_{ext}/A\ (Adh) = -P_{ext}dV$$

The work for constant external pressure where the gas (system) volume changes from V_1 to V_2 is

$$w = -\int_{V_1}^{V_2} P_{ext}dV = -P_{ext}\ (V_2 - V_1) = -P_{ext}\Delta V$$

In this form, the pressure in atmospheres and volume in liter gives a work with the units of liter–atmospheres (Latm). 1 Latm ≈ 100 J.

For a constant external pressure, P_{ex}, and a transferred heat, q, the integrated equation for the change in internal energy is

$$\Delta E = q - P_{ext}\Delta V$$

Many combinations of heat and work give the same change in internal energy. A 100 J increase in the internal energy might arise from the addition of only 100 J of heat and no work. It might also arise solely from 100 J of work done on the system – a decrease in the system volume. Various combinations of q and w, e.g., $q = 50$ and $w = 50$ will also produce the same change in internal energy. If the equation of state for the system is known, e.g., $\Delta E = C\Delta T$, the details of heat and work transfers are not needed to establish the internal energy change.

1.6 Reversible Work

Work and heat are transferred across the boundary between the system and the surroundings. Since these transfers depend on conditions in the surroundings, they are usually not state functions. An internal energy change, by contrast, occurs entirely within the system even though energy transfers are necessary for the change. For the ideal gas with equation of state

$$\Delta E = C_v\Delta T$$

the change in internal energy is determined by the temperature change of the system. The work and heat that produced the change need not be known.

Work is defined only in terms of the external pressure but some pressure gradient is required for a transfer to take place. The pressure of the gas, the internal pressure, must be larger than the external pressure if an expansion occurs. Similarly, the temperature of the gas, the internal temperature, must be larger than the external surroundings temperature if heat is to leave the system.

The theoretical limit where the temperature gradient goes to zero, i.e., $T_{int} = T_{ext}$, is one example of a reversible limit. Since both system and surroundings are at the same temperature, heat can move in or out of the system with equal facility. Heat is transferred slowly and the system remains essentially at equilibrium.

Work also has a reversible limit. A frictionless piston in a cylinder separating an internal pressure of 10 atm and an external pressure of 1 atm is not a reversible system. The piston expands irreversibly until the internal pressure reaches equilibrium at 1 atm and $P_{ext} = P_{int}$ at constant temperature. The piston does not spontaneously return to its original position since the two pressures are equal.

The work for the one-step expansion from 1 to 10 L against a constant external pressure of 1 atm at constant temperature is

$$W = 1 \int_{1}^{10} dV = -9 \, \text{Latm} = -900 \, \text{J}$$

(1 Latm $\approx$ 100 J). The irreversible expansion is also described by a piston with ten weights that combine for a total pressure of 10 atm. When nine weights are removed, the piston rises against the remaining weight (1 atm pressure).

The expansion from 1 to 10L can be accomplished in two steps. Five weights (5 atm) are removed to produce $P_{ext} = 5$ atm. The internal pressure expands the gas until $P_{ext} = P_{int} = 5$ atm and $V = 2$ L ($PV = 10$ Latm $= 2 \times 5$). The work is

$$w = -P_{ext} \int_{1}^{2} dV = -5 \, \text{atm} \, (2 - 1) \, \text{L} = -5 \, \text{Latm}$$

The expansion to 10 L occurs in a second step when the external pressure is reduced to 1 atm and the piston expands from 2 to 10 L doing work

$$w = -P_{ext} \int_{2}^{10} dV = -1 \, \text{atm} \, (10 - 2) \, \text{L} = -8 \, \text{Latm}$$

The total work for this two-step process

$$W = -5 - 8 = -13 \, \text{Latm}$$

is larger than the -9 L atm for a single-step expansion from 1 to 10 L. The smaller pressure differentials between the internal and the external pressure for the two-step process generated additional work. Work depends on the path between the initial and the final states. The system does more work when the difference between the internal and the external pressures is smaller, e.g., the two-step expansion.

A three-step expansion produces more work than the two-step expansion. For the expansion sequence,

$$(10\,\text{atm},1\,\text{L}) \rightarrow (5\,\text{atm},2\,\text{L}) \rightarrow (2\,\text{atm},5\,\text{L}) \rightarrow (1\,\text{atm},10\,\text{L})$$

the total work is $-16\,\text{L atm}$. The 3 L atm improvement in work done is less than the 4 L atm improvement in going from a one-step to a two-step expansion. The total work done reaches a limit as the number of expansion steps increases.

Maximal work results as internal and external pressures are equal for each step. In this limit, no motion is possible. However, if the external pressure is infinitesimally smaller than the internal pressure, the piston expands infinitely slowly, i.e., at equilibrium and reversibly. In this limit, $P_{\text{ext}} = P_{\text{int}}$ and the volume changes by dV. After each volume change dV, the external pressure must be reduced to keep $P_{\text{ext}} = P_{\text{int}}$.

For 1 mol of an ideal gas with

$$P_{\text{int}} = P = \frac{RT}{V}$$

the differential reversible work is

$$dw = -P_{\text{ext}}dV = -P_{\text{int}}dV = -\frac{RT}{V}dV$$

For the expansion from 1 to 10 L at constant temperature

$$dw = -RT \int_1^{10} dV/V$$

$$w_{\text{rev}} = -RT \ln V|_1^{10} = -RT \ln \frac{10}{1}$$

A change V_1 to V_2 for n mol of ideal gas at constant temperature has reversible work

$$w_{\text{rev}} = -nRT \ln \frac{V_2}{V_1}$$

Only state (system) variables (V, n, and T) appear in the final expression.

The total reversible work has units of Latm because the product $PV = nRT$ has these units. For the 1–10 L expansion, $PV = 10$ Latm and

$$w = -nRT \ln \frac{10}{1} = -10 \ln 10 = -23 \, \text{Latm}$$

This is the maximal work possible for the expansion from 1 to 10 L at constant T and n.

Weights on the piston illustrate reversibility. An infinitesimally small portion of the 10 atm is removed from the piston and stored on a platform at that level and the piston rises infinitesimally. It takes essentially no energy to move the infinitesimal weight back onto the piston and return the piston to its starting position. Small weights are removed and stored at successive levels as the piston rises and the system expands reversibly toward 10 L. At any stage, the small weight at that level can be restored to reverse the direction of expansion.

1.7 Work Cycles

The one-step expansion occurred when the external pressure was decreased from 10 to 1 atm. To restore the system to 1 L in one step, an external pressure of 10 atm is required

$$w_{\text{comp}} = -10 \, \text{atm} \, (1 - 10) = +90 \, \text{Latm}$$

The irreversible work required to restore the piston in one step is 10 times the work the system provided on expansion because the external pressure, the pressure defining work, is so large. A cycle is a sequence of changes that brings the system back to its starting point. In this case, the expansion against 1 atm followed by a compression with 10 atm returns the system to its starting point with a net $90 - 9 = +81$ Latm. One full cycle for this irreversible machine requires 81 Latm of work from the surroundings. A negative net work is needed to use this piston system as an engine.

A two-step expansion followed by a two-step compression to restore the system still requires work from the surroundings. The expansion from 1 to 2 L against the external pressure of 5 atm followed by an expansion from 2 to 10 L against an external pressure of 1 atm gives a total work of -13 Latm. For a two-step compression, the external pressure is increased from 1 to 5 atm and then to 10 atm. The volume decreases from 10 to 2 L and then from 2 to 1 L

$$w = -5 \, \text{atm} \int_{10}^{2} dV + (-10 \, \text{atm}) \int_{2}^{1} dV = 50$$

and $w_{\text{net}} = 50 - 13 = +37$ Latm. Less external work is required but the piston cycle does not produce useful work.

The three-step compression requires a net surrounding's work of $+19$ Latm. The compression work (35 Latm) and net work decrease with more steps and smaller pressure differentials.

For a reversible compression back to 1 L, the external pressure equals the internal pressure and reversible compression work for the change from 10 to 1 L is

$$w_{\text{rev}} = -\int\limits_{10}^{1} \frac{nRT}{V}dV = -nRT \ln \frac{1}{10} = -10\,(-2.3) = +23\,\text{Latm}$$

In the reversible limit, the system does 23 Latm of work on expansion (–23 Latm). The work +23 Latm by the surroundings restores the piston and completes the cycle. No usable net work by the system is produced. Net work is zero for a cycle with reversible expansion and compression.

1.8 Other Types of Work

PV work is useful for gases in cylinders. Other systems can transfer energy with different types of work or combinations of works. Each work independently contributes to changes in the internal energy of the system so that

$$dE = \delta q + \sum \delta w_i$$

Work is external force × distance

$$w = F_{\text{ext}}dx$$

where increasing x adds potential energy to a system such as a rubber band or other polymer.

As the band is stretched, the molecules in the band exert an opposing force, the tension τ. For a reversible expansion, tension replaces external force

$$dw = \tau dx$$

If the band is allowed to contract by lowering the external force, the system does work.

Changes in the surface area A of a system, e.g., a balloon, involve work. A differential change in work, dw, is directly proportional to a differential change dA through the surface tension (with units of energy/area, e.g., J/m^2)

$$w = \gamma dA$$

The work is positive as area increases if the surface is assigned to the system.

An electrochemical cell converts the energy of chemical reactions in the system to electrical energy. A potential electrical potential, ψ, with units of energy per charge, e.g., J/C appears across the electrodes and the product of P and the charge transferred dq is work

$$dw = -\psi dq$$

where positive ψ will drive positive q (and work) from the system.

Each work is the product of an intensive and extensive variable. The intensive variables like T and P are independent of amount. The temperatures of a pond and a puddle can be exactly the same even though the pond contains far more water. The pressure can be the same on a person or a large room. Extensive variables like electrical charge q, volume V, and moles vary with "extent." If q doubles, the work doubles.

Intensive variables are often identified as "per some extensive quantity," e.g., energy/mol, energy/coulomb, or force/area.

1.9 Enthalpy

Internal energy, temperature, volume, and pressure for a gaseous system are system parameters that differ from transfer quantities like work or heat. A change in a system or state variable is calculated directly as the difference of its final and initial values and is path independent. Temperature, for example, could be raised in two steps from 300 to 350 K and then from 350 to 400 K. The temperature difference is 100 K. The internal energy change for an ideal gas

$$\Delta E = C_v \Delta T = C_v (T_f - T_i)$$

is also independent of the path that led from 300 to 400 K.

A state function differential, e.g., $d(X)$, is integrated to X. The X limits are then subtracted to give the change. The state function E with differential dE integrates as

$$\int_{E_i}^{E_f} d(E) = E|_{E_i}^{E_f} = E_f - E_i$$

or

$$\Delta E = E_f - E_i$$

The work with P_{ext} is not a state function even though it seems integrable to a difference of final and final states

$$-\int_{V_i}^{V_f} d(P_{ext}V) = -P_{ext}(V_f - V_i)$$

because the external pressure is not a system variable.

A state function can be built from any combination of state variables with consistent units. The product of system pressure and volume PV has units of energy and, since P and V are system variables, this product is also a state function

$$\Delta (PV) = P_f V_f - P_i V_i$$
$$P_f V_f = RT_f$$

This energy is added to the internal energy to give a new state function H, the enthalpy

$$H = E + PV$$

The change in enthalpy depends only on final and initial values

$$\Delta H = H_f - H_i = (E_f - E_i) + (P_f V_f - P_i V_i)$$

The internal energy of 1 mol of an ideal gas is constant if the temperature is constant, $\Delta E = 0$. At a constant temperature, $P_i V_i = RT = P_f V_f$

$$\Delta (PV) = P_f V_f - P_i V_i = 0$$

and

$$\Delta H = \Delta E + \Delta (PV) = 0$$

Enthalpy changes only if the temperature changes for an ideal gas.

The enthalpy change of an ideal gas depends only on the change in temperature since

$$\Delta E = C_v \Delta T$$

and

$$P_i V_i = RT_i$$

$$\Delta H = \Delta E + \Delta (PV) = C_v \Delta T + RT_f - RT_i = [C_v + R] \Delta T$$

For an ideal gas, both C_V and R are constants. For a change dT, the differential enthalpy is

$$dH = C_v dT + R dT = (C_v + R) dT = C_P dT$$

The new heat capacity for enthalpy is the heat capacity at constant pressure, $C_p > C_v$. The enthalpy increase is larger than internal energy increase if the temperature is increased. What exactly is the extra energy? Consider a cylinder/piston system with ideal gas equilibrated so that internal and external pressures are 1 atm. As temperature increases, both pressures are maintained at 1 atm. The volume must increase to satisfy the ideal gas law. The piston moves and extra energy must be added to the system

$$\Delta H = \Delta E + \Delta PV = \Delta E + P \Delta V$$

The enthalpy change is the sum of changes in state functions E and PV with P_{int} constant. $P\Delta V$ is not work which requires P_{ext} but it is an additional system energy incorporated in the enthalpy.

Both internal energy and enthalpy changes can be calculated for any system. For the ideal gas,

$$\Delta E = C_v \Delta T \qquad \Delta H = C_p \Delta T$$

The subscripts v and P suggest that internal energy is calculated only at constant volume and H is calculated only for constant pressure. Energy and enthalpy changes are determined for any system if the initial and final E, P, V are known.

1.10 Heat Capacities as Partial Derivatives

The ideal gas is characterized by internal energy and enthalpy changes that depend only on the temperature change of the system

$$\Delta E = C_v \Delta T$$

As differentials,

$$dE = C_v dT$$

And the heat capacity is the slope (derivative)

$$C_v = \frac{dE}{dT}$$

More generally, a change in internal energy might be caused by a change in either temperature or volume (for 1 mol, pressure is dependent on V and T). Two derivatives are now required

$$dE/dT \quad dE/dV$$

A state function is independent of path; dE is determined via a path where temperature is changed, while V is held constant. Once this leg is completed, T is held constant while the volume is changed

$$dE = \left(\frac{\partial E}{\partial T}\right)_V dT + \left(\frac{\partial E}{\partial V}\right)_T dV$$

∂ is the symbol for a partial derivatives where, for example, the variable V is held constant while E is differentiated with respect to T. For example,

$$\left(\frac{\partial\,[RT/V]}{\partial T}\right)_V = \frac{R}{V}\left(\frac{\partial T}{\partial T}\right) = \frac{R}{V}$$

The formal expression for heat capacity at constant volume is

$$C_V = \left(\frac{\partial E}{\partial T}\right)_V$$

By convention, the path selected for enthalpy has separate paths for T and P

$$dH = \left(\frac{\partial H}{\partial T}\right)_P dT + \left(\frac{\partial H}{\partial P}\right)_T dP$$

The heat capacity at constant pressure is

$$C_P = \left(\frac{\partial H}{\partial T}\right)_P$$

The heat capacity C_V for

$$E = C_v VT$$

is

$$\left(\frac{\partial E}{\partial T}\right)_V = C_v V\frac{dT}{dT} = C_v V$$

if C_V is constant.

The heat capacity at constant pressure for this equation of state

$$C_P = \left(\frac{\partial H}{\partial T}\right)_P$$

uses

$$H = C_v VT + PV$$

to give

$$C_p = C_V V + P\left(\frac{\partial V}{\partial T}\right)$$

Problems

1.1 Set up the integrals for the reversible work done by gases with equations of state

$$\left(P + \frac{a}{V^2}\right)(V - b) = RT$$

$$\left(P + \frac{a}{TV^2}\right)(V - b) = RT$$

$$P(V - b)\, e^{\frac{a'}{RTV}} = RT$$

for a reversible isothermal expansion from V_1 to V_2.

1.2 n moles of an ideal gas are allowed to expand isothermally from 100 atm, 1 L to 1 atm, 100 L by three different paths: (a) reversibly; (b) by having the eternal pressure drop discontinuously from 100 to 1 atm to produce an irreversible expansion; and (c) the pressure is dropped discontinuously from 100 to 50 atm and when the gas comes to thermal equilibrium at 50 atm, the pressure again decreases discontinuously from 50 to 1 atm. Calculate the work done by the gas in each case.

1.3 One mole of a monatomic ideal gas changes state by losing 12.1 L atm of heat while 8 Latm of work is done as a piston expands against an external pressure of 2 atm. Find ΔV, ΔE, and ΔT.

1.4 Determine expressions for the work when 1 mol of gas with equation of state

$$P(V - b) = RT$$

expands from V_1 to V_2 isothermally (at constant temperature) and

a. reversibly and b. against a constant final P_{ex}.

1.5 Show the reversible work for an isothermal expansion of an ideal gas can also be written as

$$W = -RT \ln(P_2/P_1)$$

1.6 One mole of an ideal monatomic gas that is initially at 300 K and 1 atm pressure undergoes a change to increase its enthalpy by 2500 J. Determine the temperature change and the change in internal energy.

1.7 One mole of an ideal monatomic gas, initially at 300 K, is expanded against an external pressure of 2 atm. During the expansion, the internal energy E decreases by 2500 J.

a. Determine the work if $q=0$.
b. Determine ΔH.

1.8 A rubber band, which changes length rather than volume, uses work $F_{ext}dx$. If the band has an equation of state

$$\tau = k(x - x_o)$$

determine the reversible work (τdx) required to stretch this band from x_o to $2x_o$.

If the band has a heat capacity C and no heat escapes from the band during the expansion, determine the internal energy change.

Chapter 2
First Law Formalism

2.1 The Special Character of State Variables

A gas can be characterized by a set of state variables. Some, such as temperature, pressure, and volume, are measured directly in the laboratory. Others, such as internal energy and enthalpy, are determined by measuring work and heat transferred or determined from changes in the experimentally observable state variables.

Changes in state variables are determined if the initial and final values of these variables are known. For example,

$$\Delta T = T_2 - T_1$$

The actual path for the change is irrelevant. If internal energy is proportional only to temperature, the initial and final temperatures alone dictate the total temperature difference needed to calculate the change in internal energy.

Gravitational potential energy is proportional to the state variable height (h). A mass m has a potential energy mgh where g is the acceleration of gravity. Ground level is $h=0$. The energy difference between the mass m at heights h_1 and h_2,

$$\Delta E = E_2 - E_1 = mg(h_2 - h_1)$$

depends only on the initial and final heights and not the path.

Heat and work can vary when a mass is raised from $h=0$ to h at the top of a building. A student (as the system) who wishes to go from the ground floor to the top floor of the building by elevator loses little chemical energy as work and generates little heat. The sum of heat and work is mgh. A student who elects to run up the stairs loses significant energy as work while producing considerable heat so that the sum is again mgh. q and w are path dependent, whereas E is not.

A system with initial values T_1, V_1, and E_1 is changed to have final values T_2, V_2, and E_2. In a second step, the variables are returned to T_1, V_1, and E_1, the initial values. The net change in all the state variables for this cycle is 0,

$$\Delta T = T_1 - T_1 = 0$$
$$\Delta V = V_1 - V_1 = 0$$
$$\Delta E = E_1 - E_1 = 0$$

M.E. Starzak, *Energy and Entropy*, DOI 10.1007/978-0-387-77823-5_2,
© Springer Science+Business Media, LLC 2010

A transition from state 1 to state 2 with an internal energy change of +100 J will require $\Delta E = -100$ J to return to the original state by any path; the total energy change for the full cycle is 0,

$$\Delta E_{cycle} = 0 = \Delta E_1 + \Delta E_2 = +100 - 100 = 0$$

Any path, not just the actual path followed in the laboratory, is used to determine a state function. Heating a gas might produce simultaneous increases in temperature and volume. The internal energy can be calculated on a two step path:

(1) determine the change in energy with a temperature change at constant volume and
(2) hold the final temperature constant and determine the change in energy with volume at constant T

$$dE = \left(\frac{\partial E}{\partial T}\right)_V dT + \left(\frac{\partial E}{\partial V}\right)_T dV$$

2.2 Energy and Enthalpy for Chemical Reactions

An isomerization

$$A \rightarrow B$$

is a rearrangement of bonds for a set of atoms. The new bonds have different energies so energy is released or absorbed on reaction. If the bonds in the B isomer require less energy than the bonds of the A isomer, some bond energy is released during the reaction and appears as system kinetic energy of the B isomers. Since this energy is proportional to the temperature, the temperature of the system increases when this energy is released.

Internal energy changes for a reaction at constant volume are listed for a constant temperature, e.g., 25°C. However, the isomer reaction above releases energy and the temperature should rise following the reaction. If 10,000 J of energy is released by the reaction, this would appear as an internal energy increase,

$$E = +10,000\,J$$

with the concomitant temperature increase. Many reactions release considerable energy and the resultant temperature change produced by this release would lead to an unrealistic rise in the system temperature. To alleviate this problem, the chemical reaction internal energies are defined as the heat transferred to maintain the constant system temperature. Thus, the 10,000 J of heat released when the new bonds are formed defines a negative, or exothermic, internal energy of reaction

$$\Delta E_r = q_v = -10,000\,J$$

An endothermic reaction requires extra energy transferred from the surroundings. Because the system gains energy at constant temperature, an endothermic reaction is positive.

The enthalpy of a reaction such as the isomerization is determined for constant pressure conditions. The difference between the energy and the enthalpy for the same reaction is illustrated by the dissociation reaction at constant volume,

$$A \rightarrow 2B$$

where $\Delta E = -50,000\,J$. The same reaction occurring in a cylinder fitted with a frictionless piston is reversible when the pressure within the system is always equal to the external pressure on the piston so that the system pressure remains constant. The increase in moles by dissociation changes the gas volume and the piston must rise to accommodate the new volume. Some of the energy released by the reaction raises the piston so less is available to leave the system and maintain 25°C.

The enthalpy for the gaseous reaction is predicted using the enthalpy definition and the ideal gas law

$$\Delta H_r = \Delta E_r + \Delta (PV)$$
$$\Delta H = \Delta E + \Delta (nRT) = \Delta E + (\Delta n) RT$$

since the temperature is constant when this energy leaves.

The dissociation

$$A \rightarrow 2B$$

in a closed (constant volume) vessel gives an internal energy change of $-50,000$ J for the reaction as written. Since both A and B are gaseous, reaction produces a net increase of 1 mol of gas,

$$\Delta n = 2 - 1 = 1$$

and the predicted enthalpy for this reaction is

$$\Delta H = \Delta E + \Delta nRT = -50,000 + (+1)(8.31)(298) = -47,523\,J$$

The reaction under constant pressure conditions releases 2476 fewer joules, the energy required to raise the piston.

The enthalpy of reaction is defined for gases in a cylinder with constant pressure. Reactions of solids and liquid involve small changes in volume and internal energy and enthalpy are almost equal. By convention, the enthalpy of reaction is recorded for constant pressure.

2.3 Hess's Law and Reaction Cycles

The energy and enthalpy of a chemical reaction are determined only by the initial (reactants) and final (product) states and not the path the system follows to move between these states. The internal energy change for the reaction is the difference between these product and reactant energies and is independent of the reaction path between these species. The enthalpy of reaction is also path independent.

The path independence is demonstrated with a hypothetical (and incorrect) situation where the energy of reaction does depend on path. Two different reaction pathways give different energies. Reactant (A) reacts to product (B) through an intermediate C

$$A \rightarrow C \rightarrow B$$

with a total enthalpy of reaction, $\Delta H = -1000$ J for the two steps. If A goes to B directly in one step, $\Delta H = -750$ J. If the one-step reaction from A to B releases 750 J of energy, the reverse (endothermic) reaction from B to A must absorb 750 J. The two reactions are combined to form a cycle that begins and ends with A. The forward reaction follows the two-step path, while the reverse reaction to complete the cycle is the one-step path. The reactions for the full cycle are

$$A \rightarrow C \rightarrow B \ \Delta H = -1000$$
$$B \rightarrow A \ \Delta H = +750$$

The enthalpy for the cycle (H_{cyl}) is determined by adding the enthalpies of reaction for each step in the cycle to give -250 J for this hypothetical situation. A is exactly the same but 250 J of energy has been created in violation of the conservation of energy. The state function, enthalpy, for the reaction depends on the final (B) and initial (A) states. Each path between reactants and product must give exactly the same result.

The total enthalpy change for the cycle is

$$\Delta H_{cycle} = -1000 + 1000 = 0$$

The path independence of reactions is the basis of Hess' law. The unknown enthalpy or energy for a reaction is determined by combining other reactions with known energies and enthalpies.

The enthalpy required to change carbon as graphite into carbon as diamond cannot be determined directly since graphite does not change into diamond at normal temperatures and pressures. However, this enthalpy can be determined by the simple, but expensive, task of burning both graphite and diamond and determining their enthalpies for combustion

$$C\,(gra) + O_2 \rightarrow CO_2 \qquad \Delta H - 388.7\,kJ$$
$$C\,(dia) + O_2 \rightarrow CO_2 \qquad \Delta H = -391.2\,J$$

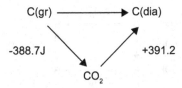

Fig. 2.1 Two paths to create graphite from diamond

The oxidation equation for diamond is reversed

$$CO_2 \rightarrow C\,(dia) + O_2 \quad \Delta H = +391.2\,J\,mol^{-1}$$

If 391.2 kJ of energy is released when diamond is burned, 391.2 kJ must be added to remake the diamond from CO_2. The enthalpy is 2.5 J for the two-step path and also the direct path from graphite to diamond.

The enthalpy for the reaction from graphite to diamond uses the two combustion reactions to complete the path from graphite to diamond (Fig. 2.1).

The two reactions can be added and their enthalpies with proper sign added to give the final reaction and enthalpy,

$$C\,(gr) + O_2 \rightarrow CO_2 \qquad \Delta H - 388.7\,kJ$$
$$CO_2 \rightarrow C\,(dia) + O_2 \qquad \Delta H = 391.2\,kJ$$
$$C\,(gr) \rightarrow C\,(dia) \qquad \Delta H = 391.2 - 288.7 = +2.5\,J$$

The graphite to diamond enthalpy is also determined by placing each reaction on a vertical scale. CO_2 is a common energy and C(dia) is 2.5 J above C(gra) on this scale (Fig. 2.2).

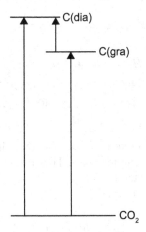

Fig. 2.2 Relative enthalpies for diamond and graphite

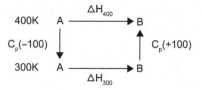

Fig. 2.3 A three-step path for reaction at 400 K when its energy at 300 K and the heat capacities are known

Any state function change can be determined using the most convenient pathway. For example, the enthalpy for the reaction,

$$A \rightarrow B$$

is $-10{,}000$ J at 300 K. The enthalpy for the same reaction at 400 K is determined with a path that goes from reactants to products at 400 K. One such path (Fig. 2.3) uses three steps: (1) A at 400 K is cooled to 300 K, (2) A is changed to B at 300 K, and (3) B at 300 K is heated to 400 K.

The change in enthalpy for a change in temperature is determined via the heat capacities at constant pressure, C_p, for $A(C_p(A))$ and B $(C_p(B))$. The enthalpy change for A when it is cooled to 300 K is

$$\Delta H = C_P(A)\,\Delta T = C_P(A)(300 - 400) = C_P(A)(-100)$$

The change from A to B at 300 K is already known to be $-10{,}000$ J. The enthalpy change for step 3 is

$$\Delta H_3 = C_P(B)\,\Delta T = C_P(B)(400 - 300) = C_P(B)(+100)$$

The enthalpy change for the reaction at 400 K is the sum of these three steps,

$$\Delta H_r\left(400^\circ\right) = \Delta H_1 + \Delta H_2 + \Delta H_{3\#}$$
$$= C_P(A)(300 - 400) + (-10{,}000) + C_P(B)(400 - 300)$$
$$= -10{,}000 + (C_P(B) - C_P(A))(400 - 300)$$
$$= -10{,}000 + \Delta C_P \Delta T = -10{,}000 + \Delta C_P(100)$$

where the ΔC_p is just the difference between the heat capacities of all the products less the heat capacities of all the reactants. The second term corrects the reaction enthalpy at 300 K for the new temperature.

In general, the enthalpy for a reaction,

$$aA + bB \leftrightarrow cC$$

at any temperature includes the net difference of heat capacities for the reactants and products,

$$\Delta C_P = cC_P(C) - aC_P(A) - bC_P(B)$$

When the product and reactant total heat capacities are similar, the enthalpy changes little with temperature and is essentially constant. This property is useful when a constant enthalpy gives a linear plot in some graphs.

2.4 Standard States

The CO_2 enthalpy served as a common energy for both graphite and diamond. When both had a common reference, their energy of reaction was the difference between their energies. Standard states use this information to select an energy that is common to all molecules as a reference or standard state. The reference is the energy of the atoms of the molecules in their most common states. Enthalpy or energy differences for the molecules are then differences of these standard state enthalpies. For the graphite–diamond example, $C(gra)$ is selected as the standard state and assigned a standard enthalpy 0. $C(dia)$ is then 2.5 J mol^{-1} higher so its standard enthalpy is 2.5 J mol^{-1}. The CO_2 molecule has enthalpy 388.7 kJ less than $C(gr)$ so its standard enthalpy is -388.7 kJ. The enthalpy to convert diamond into CO_2 is just the difference of standard state enthalpies for product CO_2 (-388.7) and the reactant diamond (2.5)

$$\Delta H = H_f^\circ(CO_2) - H_f^\circ(C\,(dia)) = -388.7 - 2.5 = 391.2\,J\,mol^{-1}$$

The subscript f indicates formation from the elements, while the o indicates 25°C. Deltas are absent because the standard states are all relative to an assigned value of 0 for the elements in their standard states.

For a general reaction,

$$aA + bB \leftrightarrow cC + dD$$

the total enthalpy of formation for products is the sum of the enthalpies of formation of all the products less the enthalpies of formation of all the reactants

$$[cH_f^\circ(C) + dH_f^\circ(D)] - [aH_f^\circ(A) + bH_f^\circ(B)]$$

For a reaction,

$$A + 2B = 2C$$

the enthalpy of reaction is

$$\Delta H_r = 2\Delta_f^\circ(C) - 1\Delta H_f^\circ(A) - 2\Delta H_f^\circ(B)$$

The negative signs for the reactants show that these reactions have been reversed to produce the proper net reaction.

2.5 More Partial Derivatives

Partial derivatives are used to establish a path involving several independent variables. Each term is a slope (partial derivative) and a differential. The power of the formalism is revealed when the partial derivatives are known.

The pressure P of an ideal gas is related to the volume and temperature as the independent variables so that

$$dP = \left(\frac{\partial P}{\partial T}\right)_V dT + \left(\frac{\partial P}{\partial V}\right)_T dV$$

A differential dP is determined by the derivatives and differentials for T and V. The derivatives are evaluated from ideal gas law

$$P = RT/V$$

The change in P with T at constant V is

$$\left(\frac{\partial P}{\partial T}\right)_v = \left(\frac{\partial \left(\frac{RT}{V}\right)}{\partial T}\right)_v = \frac{R}{V}\left(\frac{\partial T}{\partial T}\right) = \frac{R}{V}$$

and the change in P with V at constant T is

$$\left(\frac{\partial P}{\partial V}\right)_T = \left(\frac{\partial \left(\frac{RT}{V}\right)}{\partial V}\right)_T = RT\left(\frac{\partial \left(\frac{1}{V}\right)}{\partial V}\right)_T = RT\left(-\frac{1}{V^2}\right)$$

to generate dP for an ideal gas

$$dP = \frac{R}{V}dT - \frac{RT}{V^2}dV$$

The ideal gas constant R is approximated as 0.08 Latm K^{-1} mol^{-1} to give an initial pressure, volume, and temperature $(P, V, T) = (1$ atm, 10 L, 125 K$)$ for 1 mol of gas. The final (P, V, T) are (2 atm, 20 L, 500 K). The pressure change is

$$dP = \int_1^2 dP = 2 - 1 = 1$$

Using the right side of the equation, T is changed from 125 to 500 K with $V = 10$ L. The constant temperature is 500 K during the second leg. The pressure change for the first leg

$$\Delta P = (+ R/V)\Delta T = +0.08/(10 \text{ L}) [500 - 125] = +3$$

is added to the pressure change for the second leg

$$\Delta P = -RT \int_{10}^{20} \frac{dV}{V^2} = +\frac{RT}{V} \Big|_{10}^{20} = \frac{RT}{V_2} - \frac{RT}{V_1}$$

$$0.08 \, (500) \left(\frac{1}{20} - \frac{1}{10} \right) = 40 \, (0.05 - 0.1) = -2 \, \text{atm}$$

The net change in pressure for the two-step path is $3\text{--}2 = 1$. If volume is changed first while $T=125$, the result is the same

$$\Delta P = RT \int_{10}^{20} -\frac{1}{V^2} dV + \frac{R}{V} \int_{125}^{500} dT = (0.08) \, (125) \, \frac{1}{V} \Big|_{10}^{20} + \frac{0.08}{20} T \Big|_{125}^{500}$$
$$= 10 \, (0.05 - 0.1) + 0.004 \, (375) = 1$$

Path differential equations are also used for error analysis. If the experimental errors in measuring T and V are ΔT and ΔV, respectively, each of these errors could produce an error in the value of P, ΔP. The rates of change of pressure with respect to T and V are multiplied by the errors ΔT and ΔV, respectively

$$BP = \frac{R}{V} \Delta T$$

$$\Delta P = -\frac{RT}{V^2} \Delta V$$

These independent errors are combined by squaring each, adding, and then taking the square root of the sum

$$\Delta P^2 = \sqrt{\left[\left(\frac{\partial P}{\partial T} \right)_V dT \right]^2 + \left[\left(\frac{\partial P}{\partial V} \right)_T dV \right]^2}$$

The change in internal energy is expressed as the sum of T and V legs:

$$dE = C_v dT + \left(\frac{\partial E}{\partial V} \right)_T dV$$

The derivatives for some simple systems are 0 and this simplifies the path expressions. For example, an ideal gas has

$$\left(\frac{\partial E}{\partial V} \right)_T = 0$$

so that

$$dE = C_v dT + \left(\frac{\partial E}{\partial V} \right)_T dV = C_v dT$$

2.6 Generalized Thermodynamic Equations

The relation between the heat capacity at constant pressure and the heat capacity at constant volume

$$C_P = C_V + R$$

is valid for an ideal gas. A more general expression works for any equation of state with P, T, and V.

The heat capacity at constant pressure is

$$\left(\frac{\partial H}{\partial T} \right)_P$$

Since

$$H = E + PV$$

the terms on the right side of the equation are differentiated with respect to T at constant pressure. The derivative of H with respect to T at constant P is

$$\left(\frac{\partial H}{\partial T} \right)_P = \left(\frac{\partial E}{\partial T} \right)_P + P \left(\frac{\partial V}{\partial T} \right)_P$$

The partial derivative

$$\left(\frac{\partial E}{\partial T} \right)_P$$

is not a standard heat capacity and must be related to C_V.

The internal energy is expressed in terms of its standard variables, T and V, using partial derivatives:

$$dE = \left(\frac{\partial E}{\partial T} \right)_V dT + \left(\frac{\partial E}{\partial V} \right)_T dV$$

These T and V differentials become partial derivatives if they are each differentiated with respect to T at constant P:

$$\left(\frac{\partial E}{\partial T} \right)_P = \left(\frac{\partial E}{\partial T} \right)_V \left(\frac{\partial T}{\partial T} \right)_P + \left(\frac{\partial E}{\partial V} \right)_T \left(\frac{\partial V}{\partial T} \right)_P = C_V + \left(\frac{\partial E}{\partial V} \right)_T \left(\frac{\partial V}{\partial T} \right)_P$$

Substituting this equation into the heat capacity expression to gives the general expression relating C_P and C_V,

$$C_P = \left(\frac{\partial H}{\partial T}\right)_P = C_V + \left(\frac{\partial E}{\partial V}\right)_T \left(\frac{\partial V}{\partial T}\right)_P + P\left(\frac{\partial V}{\partial T}\right)_P$$

$$\#C_P = C_V + \left[\left(\frac{\partial E}{\partial V}\right)_T + P\right]\left(\frac{\partial V}{\partial T}\right)_P$$

for any gas with any equation of state. For an ideal gas, with

$$\left(\frac{\partial E}{\partial V}\right)_T = 0$$

and

$$\left(\frac{\partial V}{\partial T}\right)_P = \frac{R\,dT}{P\,dT} = \frac{R}{P}$$

$$C_P = C_V + [0 + P]\frac{R}{P} = C_V + R$$

as expected.

A dependent state variable that depends on two (or more) other thermodynamic variables might be constant during a change of state so that its differential is 0 and the partial derivative terms on the right side of the equation must also equal 0. For example, if the enthalpy remains constant during an expansion, $dH = 0$ and

$$dH = 0 = \left(\frac{\partial H}{\partial T}\right)_P dT + \left(\frac{\partial H}{\partial P}\right)_T dP$$

For the independent variables, T and P, the differentials (not the partials) are differentiated with respect to temperature to give

$$\left(\frac{\partial H}{\partial T}\right)_P \frac{dT}{dP} = -\left(\frac{\partial H}{\partial P}\right)_T \left(\frac{dP}{dP}\right)$$

$$\left(\frac{\partial H}{\partial T}\right)_P \frac{dT}{dP} = -\left(\frac{\partial H}{\partial P}\right)_T$$

The standard derivative dT/dP is replaced with partial derivatives to show the process proceeds at constant enthalpy:

$$\frac{dT}{dP} = \left(\frac{\partial T}{\partial P}\right)_H = -\frac{\left(\frac{\partial H}{\partial P}\right)_T}{\left(\frac{\partial H}{\partial T}\right)_P}$$

2.7 Calculating Internal Energy

The differential expansion for dE

$$dE = \left(\frac{\partial E}{\partial T}\right)_V dT + \left(\frac{\partial E}{\partial V}\right)_T dV$$

has derivatives that depend on the specific system under study. For example, an ideal gas has a constant heat capacity:

$$\left(\frac{\partial E}{\partial T}\right)_V = C_V = \text{constant}$$

and

$$\left(\frac{\partial E}{\partial V}\right)_T = 0$$

ΔE is determined by integrating dT from its initial (T_i) to its final temperature (T_f). If C_V is temperature dependent,

If

$$\left(\frac{\partial E}{\partial V}\right)_T = 0$$

$$\Delta E = \int_{T_i}^{T_f} C_V(T)\, dT$$

If $C_V = \text{constant}$,

$$\Delta E = \int_{T_i}^{T_f} C_V dT = C_V \Delta T$$

A molar heat capacity linear in temperature

$$C(T) = C_v + aT$$

gives

$$\Delta E = \int_{T_i}^{T_f} [C_v + aT]\, dT = C_v T|_{T_i}^{T_f} + \frac{aT^2}{2}|_{T_i}^{T_f} = C_v \Delta T + \frac{a}{2}[T_f^2 - T_i^2]$$

The temperature dependent term is usually small relative to the constant C_V term. For real gases, the $C_v dT$ changes are usually much larger than

$$\left(\frac{\partial E}{\partial V}\right)_T dV$$

changes. For isothermal ($dT{=}0$) conditions

$$dE = \left(\frac{\partial E}{\partial V}\right)_T dV$$

This energy change is larger for a small volume where the gas molecules are closer together on average so that intermolecular interaction energies are significant. This potential energy of interaction is released as the volume increases and the molecules separate.

One mole of a van der Waals gas has a partial derivative in volume that depends on the molecular interaction constant a,

$$\left(\frac{\partial E}{\partial V}\right)_V = \frac{a}{V^2}$$

the pressure correction term in the van der Waals equation,

$$\left(P + \frac{a}{V^2}\right)(V - b) = RT$$

with units of pressure.

If the van der Waals gas is expanded from V_i to V_f at constant temperature, $dT = 0$, and

$$dE = \left(\frac{dE}{dV}\right)_T dV$$

is integrated to give

$$\Delta E = \int_{u_i}^{u_f} dE = \int_{V_i}^{V_f} \left(\frac{\partial E}{\partial V}\right)_T dV = \int_{V_i}^{V_f} \frac{a}{V^2} dV = -\frac{a}{V}\Big|_{V_i}^{V_f} = -\frac{a}{V_f} + \frac{a}{V_i}$$

As volume increases, the internal energy will increase reflecting the change of potential energy of interaction to kinetic or internal energy as the molecules separate.

Problems

2.1 The heat of formation of HBr(g) from H_2(g) and Br_2(g) is $-$ 38 kJ mol^{-1}. Determine the heat of formation at 400 K.

2.2 A diatomic gas expands isothermally (constant temperature) against an external pressure $P_{ex} = 1$ from $V_1 = 1$ L to $V_2 = 2$ L. The gas is non-ideal and

$$\int \frac{\partial E}{\partial V} dV = -0.1 \, \text{Latm}$$

Determine w, ΔE, and q.

2.3 An ideal monatomic gas expands reversibly in two steps: (1) an isothermal ($dT = 0$) expansion at 400 K and (2) an adiabatic expansion ($q = 0$). During the two steps, the system absorbs 5000 J of heat and its internal energy decreases by -500 J mol^{-1}.

a. Determine the internal energy changes for each step.
b. Determine the work for each step.

2.4 A non-ideal monatomic gas has

$$\int_{V_1}^{V_2} \frac{\partial E}{\partial V} dV = a = \text{constant}$$

Write an integrated expression for the change in internal energy when the temperature increases from T_1 to T_2 and the volume increases from V_1 to V_2.

2.5 The standard enthalpies of formation use elements in their most stable states. For the reaction

$$A \rightarrow B$$

with

$$\Delta H_f^\circ (A) = -1000 \, \text{J}$$
$$\Delta H_f^\circ (B) = -2000 \, \text{J}$$

Determine the "standard" enthalpy of A is B is selected as the 0 enthalpy reference state.

2.6 One mole of a non-ideal gas with $C_V = 12.5 \, \text{J K}^{-1} \, \text{mol}^{-1}$ and $\left(\frac{\partial E}{\partial V}\right)_T = 20 \, \text{J L}^{-1}$ expands against a constant pressure of 0.2 atm from 1 to 11 L while the temperature decreases from 350 to 300 K.

a. Determine ΔE and w.
b. Determine ΔH if

$$\left(\frac{\partial H}{\partial P}\right)_T = 0$$

2.7 A non-ideal monatomic gas expands with $\Delta H = +1000$ J.

$$\left(\frac{\partial H}{\partial P}\right)_T = 0$$

a. Determine the final T.
b. Determine ΔE.

Chapter 3
First Law of Thermodynamics: Applications

3.1 General Equation for the First Law

ΔE is determined from changes in other state variables such as volume and temperature or by measuring the net heat and work transferred across the system boundaries.
Since both methods give the same ΔE, they are combined in a single equation:

$$dE = \left(\frac{\partial E}{\partial T} \right)_V dT + \left(\frac{\partial E}{\partial V} \right)_T dV = dq + dw_{PV} + dw_{\text{other}}$$

Terms in the equation are eliminated for different experimental conditions.
An isothermal change, i.e., constant T, gives $dT = 0$

$$\left(\frac{\partial E}{\partial T} \right)_V dT = C_V dT = 0$$

$$\left(\frac{\partial E}{\partial V} \right)_T = 0$$

for an ideal gas so

$$dE = 0 = dq + dw_{PV} + dw_{other}$$

and

$$dq = -dw_{PV} - dw_{\text{other}}$$

Any heat transferred must be compensated by an equal and opposite transfer of
work to maintain constant internal energy.

A system can be insulated to prevent the flow of heat in either direction. This is
an adiabatic system with $q = 0$. The general equation reduces to

$$dE = \left(\frac{\partial E}{\partial T} \right)_V dT + \left(\frac{\partial E}{\partial V} \right)_T dV = dw_{PV} + dw_{\text{other}}$$

M.E. Starzak, *Energy and Entropy*, DOI 10.1007/978-0-387-77823-5_3,
© Springer Science+Business Media, LLC 2010

For an ideal gas with constant heat capacity, $(\partial E/\partial V)_T = 0$ and only PV work

$$dE = C_v dT = -P_{ext} dV$$

3.2 The Joule Experiment

dE for a gas is found from changes in the independent variables T and V:

$$dE = C_V dT + \left(\frac{\partial E}{\partial V}\right)_T dV$$

Joule studied $(\partial E/\partial V)_T$ experimentally by observing gas expansion. Gas expanded from one glass bulb into a second bulb when the stopcock was opened and Joule used a thermometer to observe a temperature change on expansion (Fig. 3.1).

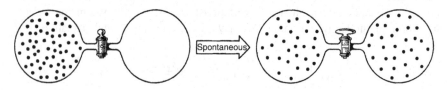

Fig. 3.1 Expansion of gas into a vacuum

Both bulbs (the system) are insulated (q=0). No work is done because the gas expands against the gas in the filling bulb, not against an external pressure. Since q=w=0, dE=0

$$dE = dq + dw = 0 + 0 = 0$$

Since V and T are the observables

$$dE = \left(\frac{\partial E}{\partial T}\right)_V dT + \left(\frac{\partial E}{\partial V}\right)_T dV = 0$$

$$C_V dT = -\left(\frac{\partial E}{\partial V}\right)_T dV$$

$$dT/dV = \frac{\frac{\partial E}{\partial V}}{C_V}$$

Since $C_v \neq 0$, *any temperature* change observed on expansion means a finite energy change with volume.

$$\left(\frac{\partial E}{\partial V}\right)_T$$

is small for the common gases. The temperature change in Joule's experiment was too small to detect with his thermometer.

3.3 The Joule–Thomson Experiment

Joule and Thomson used a different experiment to determine if enthalpy changes with pressure. The pressure differential is larger than the volume differential and any temperature change is larger. Joule and Thomson divided a hollow cylinder into two sections by inserting a porous meerschaum plug for the slow (equilibrium) transfer of gas.

Gas is transferred by applying a larger external pressure to a piston on one side so that gas moved slowly through the plug and expanded the permeant gas against a second piston. The entire cylinder is insulated, i.e., adiabatic.

The left cylinder (side 1) has an internal and applied pressure P_1 and volume V_1 initially (Fig. 3.2). When the gas expands through the porous plug, this volume drops to 0 while the right cylinder gas expands from $V_2 = 0$ to V_2 against a pressure P_2.

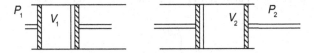

Fig. 3.2 The initial and final states for the Joule–Thomson experiment

$P_{int} = P_{ext}$ on either side and $q = 0$ and

$$\Delta E = E_2 - E_1$$

The compression of the piston on side 1 does work on the system

$$w = -P_{ext}(0 - V_1) = P_1 V_1$$

while the second piston does work

$$w = -P_{ext}(V_2 - 0) = -P_2 V_2$$

and

$$w_{net} = P_1 V_1 - P_2 V_2$$

Since both chambers constitute the system:

$$\Delta E = E_2 - E_1 = P_1 V_1 - P_2 V_2$$

Collecting terms of common subscript gives

$$E_1 + P_1 V_1 = E_2 + P_2 V_2$$

or

$$H_1 = H_2$$

Enthalpy is constant throughout the transfer and $dH = 0$

Since the experimental variables are T and P

$$dH = 0 = \left(\frac{\partial H}{\partial T}\right)_P dT + \left(\frac{\partial H}{\partial P}\right)_T dP$$

Differentiating with P

$$0 = \left(\frac{\partial H}{\partial T}\right)_P \frac{dT}{dP} + \left(\frac{\partial H}{\partial P}\right)_T \frac{dP}{dP}$$

$$\frac{dT}{dP} == \left(\frac{\partial T}{\partial P}\right)_H = -\frac{\left(\frac{\partial H}{\partial P}\right)_T}{\left(\frac{\partial H}{\partial T}\right)_P} = -\frac{\left(\frac{\partial H}{\partial P}\right)_T}{C_P}$$

is non-zero. An observed change in temperature reflects a non-zero

$$\left(\frac{\partial H}{\partial P}\right)_T$$

3.4 The Joule–Thomson Coefficient

dT/dP is the Joule–Thomson coefficient,

$$\mu_{JT} = \left(\frac{dT}{dP}\right)_H$$

Heat capacity is always positive but the Joule–Thomson coefficient and

$$\left(\frac{\partial H}{\partial P}\right)_T$$

can be positive or negative. For most gases at room temperature, cooling occurs (dT (−)) when the gas expands from a high pressure to a low pressure (dP (−)). $dT/dP > 0$. Compressed air from a gas cylinder at high pressure released to the atmosphere cools as it leaves the nozzle. Gas in a tank of "liquid" nitrogen has sufficient pressure difference to condense the gas.

Joule–Thomson cooling can liquefy most gases by reducing the temperature in a sequence of steps. The gas is compressed isothermally and then expanded to produce cooled gas. This cooled gas cools more compressed gas that is then expanded for further cooling.

Helium and hydrogen have negative Joule–Thomson coefficients at room temperature so they heat on expansion. However, the Joule–Thomsen coefficient changes from negative to positive at the Joule–Thomson inversion temperature. Hydrogen and helium are liquefied by cooling them below their Joule–Thomson inversion temperature using the expansion of other gases. They can then be cooled further by their own expansion and cooling.

3.5 The Reversible Isothermal Expansion

The general expression of the first law of thermodynamics

$$dE = C_V dT + \left(\frac{\partial E}{\partial V}\right)_T dV = dq - P_{ext}dV + dw_{other}$$

is simplified for an ideal gas with

$$\left(\frac{\partial E}{\partial V}\right)_T = 0$$

For an isothermal expansion with $dT=0$,

$$dE = 0.$$

The enthalpy expansion in T and P

$$dH = C_P dT + \left(\frac{\partial H}{\partial P}\right)_T dP = dq_P + VdP$$

simplifies to $dH = 0$ for an ideal gas since $dT = 0$ and

$$\left(\frac{\partial H}{\partial P}\right)_T = 0.$$

Since $dE=0$, the work done when the volume changes must be balanced by heat transferred across the boundary:

$$dE = 0 = dq - P_{ext}dV$$

For a reversible, ideal gas expansion

$$P_{int} = P_{ext}$$

and

$$dq = -dw = \frac{RT}{v} dv.$$

For an expansion, work is done by the system while an equal amount of heat is transferred into the system to maintain constant internal energy. The reversible, isothermal expansion of an ideal gas at 300 K from 1 to 10 L gives heat and work:

$$q_{rev} = -w_{rev} = \left(8.31 \frac{J}{K\,mol}\right)(300\,K)\ln\left(\frac{10}{1}\right) \approx 2500\,(2.3) = 5750\frac{J}{mol}$$

where w and q are calculated from the volume change.

Since state functions depend only on the initial and final thermodynamic parameters, the enthalpy for an ideal gas has

$$\left(\frac{\partial H}{\partial P}\right)_T = 0$$

and

$$dH = C_p dT + \left(\frac{\partial H}{\partial P}\right)_T dP = C_p(0) + (0)\,dP = 0$$

For $\Delta.H = 0$, absorbed heat is equal and opposite to the energy, VdP,

$$dq = -VdP$$

Since both P and V in H are system variables, V is expressed in terms of P using the ideal gas law

$$P_i = \frac{RT}{V_i}$$

to give

$$q = -RT \ln\left(\frac{\frac{RT}{V_2}}{\frac{RT}{V_1}}\right) = -RT \ln\left(\frac{V_1}{V_2}\right) = +RT \ln\left(\frac{V_2}{V_1}\right)$$

This heat equals the reversible work because it describes the same reversible transition.

3.6 Irreversible Isothermal Expansion of an Ideal Gas

A reversible expansion always gives the maximal work for a given change of volume. If an ideal gas expands irreversibly and isothermally against a constant external pressure from V_1 to V_2 at constant temperature, less work is done. However, since $dT = 0$ and the gas is ideal, the internal energy change during the expansion is 0,

$$\Delta E = C_v dT + \left(\frac{\partial E}{\partial V}\right)_T dV = 0$$

Any work done by the system during the expansion is compensated by the absorption of an equal amount of heat to maintain the constant internal energy.

The heat absorbed during the volume increase equals the work that is done against the constant external pressure:

$$q = \int dq = P_{ext} \int_{V_1}^{V_2} dV = P_{ext}(V_2 - V_1)$$

The enthalpy change is ΔE plus

$$\Delta (PV) = \Delta (RT) = R\Delta T = 0$$

and $\Delta H = 0$.

An ideal gas initially at 1 L and 24.6 atm expands isothermally (300 K) against an external pressure of 2.46 atm until its pressure reaches 2.46 atm and 10 L:

$$P = \frac{RT}{V} = \frac{\left(0.082\frac{\text{Latm}}{\text{K mol}}\right)(300\,\text{K})}{10\,\text{L}} = 2.46\,\text{atm}$$

The heat and work

$$q = -w = P_{\text{ext}}(V_2 - V_1)$$
$$= 2.46(10 - 1) \approx 22\,\text{Latm} \approx 2200\,\text{J}$$

are smaller than those for the reversible expansion.

3.7 Isothermal Expansion of Non-ideal Gases

The reversible, isothermal expansion of a gas with an equation of state for 1 mol, finite volume (b) and no intermolecular interactions ($a = 0$)

$$P(V - b) = RT$$

has

$$\left(\frac{\partial E}{\partial V}\right)_T = 0$$

and

$$\left(\frac{\partial H}{\partial P}\right)_T = b = \text{constant}$$

For an isothermal expansion ($dT = 0$) from P_1 to P_2, the differential enthalpy change

$$dH = C_p dT + \left(\frac{\partial H}{\partial P}\right)_T dP$$

gives

$$\Delta H = \int_{P_1}^{P_2} b\,dP = b(P_2 - P_1)$$

Since $\left(\frac{\partial E}{\partial V}\right)_T = 0$, the net change in internal energy for this isothermal expansion is

$$dE = C_v(0) + 0dV = 0$$

The equation of state for the van der Waals gas

$$\left(P + \frac{a}{V^2}\right)(V - b) = RT$$

has

$$\left(\frac{\partial E}{\partial V}\right)_T = \frac{a}{V^2}$$

and

$$\left(\frac{\partial H}{\partial P}\right)_T = 0$$

For an isothermal expansion,

$$dE = C_v(0) + \frac{a}{V^2}dV$$

The change in internal energy for an expansion from V_1 to V_2 is

$$\Delta E = \int_{V_1}^{V_2} \frac{a}{V^2}dV$$

$$= -a(1/V_2 - 1/V_1)$$

Since $V_2 > V_1$, $\Delta E > 0$. As the molecules separate, the potential energy of interaction is released as kinetic energy.

The enthalpy change is

$$\Delta H = b(P_2 - P_1)$$

The internal pressure for reversible work of expansion for a van der Waals gas with

$$P = \frac{RT}{V - b} - \frac{a}{V^2}$$

gives

$$w = -\int_{V_1}^{V_2}\left[\frac{RT}{V - b} - \frac{a}{V^2}\right]dV = -RT\ln(V - b)\Big|_{V_1}^{V_2} \frac{a}{V}\Big|_{V_1}^{V_2}$$

$$= -RT\ln\left(\frac{V_2 - b}{V_1 - b}\right) - \frac{a}{V_2} + \frac{a}{V_1}$$

The second term is the energy needed to separate the molecules when V increases. The equation shows that this released energy can be used to do work.

The heat is determined by combining internal energy and work

$$\Delta E = - - \frac{a}{V_2} + \frac{a}{V_1} = q - RT \ln\left(\frac{V_2 - b}{V_1 - b}\right) - \frac{a}{V_2} + \frac{a}{V_1}$$

$$q = RT \ln\left(\frac{V_2 - b}{V_1 - b}\right)$$

3.8 Adiabatic Reversible Expansions

An adiabatic system ($q{=}0$) equates internal energy and work. If a monatomic ideal gas is allowed to expand reversibly and adiabatically from a volume V_1 to a volume V_2, the work done equals the internal energy change in the system. Since temperature changes, T must be eliminated from the work expression.

The general first law equation,

$$dE = C_v dT + \left(\frac{\partial E}{\partial V}\right)_T dV = dq - P_{ext} dV$$

simplifies as follows:

(1) Since the process is adiabatic, $dq = 0$;
(2) Since the gas is ideal:

$$\left(\frac{\partial E}{\partial V}\right)_T = 0$$

(3) Since the expansion is reversible:

$$P_{ext} = P_i = P = \frac{nRT}{V}$$

for n moles of gas in the system.

(4) n moles of the ideal monatomic gas have a heat capacity $C_v = n3R/2$.

The remaining terms are

$$C_v dT = - (RT/V)\, dV$$

For integration, both sides are divided by T; the left side is integrated over T while the right side is integrated over V with the common format

$$dE = nC_v dT = \frac{n3R}{2} dT = dq - P_{ext} dV = -\frac{nRT}{V} dV$$

$$\frac{3R}{2} \int_{T_1}^{T_2} \frac{dT}{T} = - \int_{V_1}^{V_2} \frac{R}{V} dV$$

$$3/2 \ln(T_2/T_1) = - \ln(V_2/V_1)$$

For a reversible expansion from 1 to 10 L when the gas is initially at 300 K, the final temperature, T_2, is

$$\frac{3}{2} \ln \left(\frac{T_2}{300} \right) = - \ln \left(\frac{V_2}{V_1} \right) = - \ln \left(\frac{10}{1} \right)$$

$$T_2 = 300e^{-1.53} = 65 \text{ K}$$

Since the initial and final temperatures are known, the internal energy change and the reversible work are

$$\Delta E = \frac{3R}{2} (65 - 300) = -2908.5 \text{ J} = w_{\text{rev}}$$

for this reversible adiabatic expansion. $R = 8.31 \text{ J K}^{-1} \text{ mol}^{-1}$.

The initial and final temperatures give the enthalpy change as well. The heat capacity at constant pressure for this gas is

$$C_p = C_v + R = \frac{5R}{2} \approx 21 \text{ J K}^{-1} \text{mol}^{-1}$$

and

$$\Delta H = C_p (T_f - T_i) = 21 (65 - 300) = -4935 \text{ J K}^{-1} \text{mol}^{-1}$$

Because temperature, pressure, and volume are all interrelated through the ideal gas law, this expression also defines the change of temperature with pressure for a reversible adiabatic expansion. The initial and final volumes

$$V_i = \frac{RT_i}{P_i}$$

$$V_f = RT_f/P_f$$

give

$$C_v \ln \left(\frac{T_2}{T_1} \right) = -R \ln \left(\frac{\frac{RT_2}{P_2}}{\frac{RT_1}{P_1}} \right)$$

which is rearranged to

$$C_v \ln \left(\frac{T_2}{T_1} \right) = -R \ln \left(\frac{T_2}{T_1} \right) - R \ln \left(\frac{P_1}{P_2} \right)$$

$$\ln \left(\frac{T_2}{T_1} \right) = \ln \left(\frac{P_2}{P_1} \right)^{\frac{R}{C_p}}$$

The logarithms on both sides cancel to give

$$(T_2/T_1) = (P_2/P_1)^{R/C_v}$$

3.9 Irreversible Adiabatic Expansion

An adiabatic irreversible expansion of an ideal gas against a constant external pressure obeys

$$dE = C_v dT = -P_{ext} dV$$

External pressure is independent of system temperature so that both sides integrate over their separate variables (T and V).

For an initial temperature of 300 K and an expansion from 1 to 10 L against an external pressure of 1 atm, the final temperature is calculated from

$$\Delta E = C_V \int_{300}^{T_2} dT = -P_{ext} \int_1^{10} dV \frac{3}{2} R (T_2 - 300) = -1 \, \text{atm} \, (10 - 1)$$

$$= -9 \, \text{Latm} = -900 \, \text{J}$$

$$\#12.5 \, (T_2 - 300) = -900 \quad T_2 = 227 \, K$$

The internal energy change for any adiabatic expansion is equal to the work. The final temperature is determined from the change in internal energy.

For non-ideal gases, when the partial derivative

$$\left(\frac{\partial E}{\partial V} \right)_T$$

is finite

$$\int_{T_1}^{T_2} C_V dT + \int_{V_1}^{V_2} \left(\frac{\partial E}{\partial V} \right)_T dV = -P_{ext} \int_{V_1}^{V_2} dV$$

The temperature change is determined only when the two terms in volume are known.

Problems

3.1 An ideal monatomic gas expands from a volume of 2 L in the left compartment to a volume of 5 L in the right compartment in a Joule–Thomson experiment. The system is insulated. The pressure of the gas in the left compartment is $P_{ext} = P_{int} = 5$ atm, while the pressure on the gas in the right compartment is $P_{ext} = P_{int} = 2$ atm. The initial temperature is 300 K.

a. Determine dT/dP for this gas.
b. Determine ΔE

3.2 One mole of an ideal monatomic gas which is initially at 300 K and 1 atm pressure undergoes a reversible change which increases its enthalpy by 4000 J.

Determine the temperature change and the change in internal energy ΔE for the gas in this change.

3.3 One mole of a monatomic gas with the equation of state,

$$P(V - b) = RT$$

initially at 300 K, expands against a constant external pressure of 1 atm from 1 to 10 L and does irreversible work of 9 Latm = 907 J. During this process, 347 J of heat are also lost from the system.

a. The thermodynamic equation of state for this system will give $(\partial E/\partial V)_T = 0$. Determine the internal energy change and the temperature change for this system if $C_v = 3R/2 = 12.5$ J K^{-1} mol^{-1}. 1 Latm = 100 J.
b. If $b = 0.05$ J atm^{-1} for this gas, determine the enthalpy change for the change in (a). Do not forget $(\partial H/\partial P)_T = b$. Use P_f for your final pressure.

3.4 One mole of an ideal monatomic gas, initially at 300 K, is expanded irreversibly and adiabatically against an external pressure of 2 atm. During the expansion, the internal energy E decreases by 10,000 J.

a. Determine the work done by the expansion.
b. Determine ΔH and q for the expansion.
c. Determine the final temperature of the system.

3.5 Use the definition $H = E + PV$ to prove

a. $(\partial H/\partial T)_P = (\partial E/\partial \dot{T})_V + \left[P + (\partial E/\partial V)_T\right](\partial V/\partial T)_P$
b. $(\partial E/\partial T)_V = (\partial H/\partial T)_P - \left[-(\partial H/\partial T)_P (\partial T/\partial P)_H + V\right](\partial P/\partial T)_P$

3.6 Show that

$$\alpha = (1/V)(\partial V/\partial T)_P = 1/T \quad \text{and} \quad \beta = -(1/V)(\partial V/\partial P)_T = 1/P$$

for an ideal gas.

3.7 Calculate the temperature increase and final pressure of a monatomic ideal gas if 1 mol is compressed adiabatically and reversibly from 44.8 L at 300 K to 22.4 L. $C_v = 12.5$ J K^{-1} mol^{-1}.

3.8 For an energy

$$D = H + RT$$

a. Explain why D is a state function.

b. Show

$$C_P = \left(\frac{\partial D}{\partial T}\right)_P - R$$

c. Find

$$\left(\frac{\partial T}{\partial P}\right)_D$$

in terms of partials in D.

Chapter 4
Entropy and the Second Law: Thermodynamics Viewpoint

4.1 Temperature Gradients and Net Work

A piston expanding reversibly and isothermally from a volume V_1 to a final volume V_2 generates the maximal possible work for an ideal gas at constant temperature

$$w_{rev} = RT \ln \left(\frac{V_2}{V_1} \right)$$

A reversible isothermal compression from V_2 to V_1 at this temperature is $-w_{rev}$ and the work for the cycle from V_1 to V_1 is 0

$$w = -RT \ln \left(\frac{V_1}{V_1} \right) = -RT \ln (1) = 0$$

Each reversible step in the cycle gives maximal work but no net work is done for the cycle. All work generated on expansion must be stored and used to return the gas to its initial volume.

Net work is produced in a reversible cycle if the gas is recompressed at a lower temperature. Less work is then required to compress the gas and the residual work can be used externally. Each cycle then generates a net work.

An ideal gas expanding reversibly at 122 K transfers +23 Latm of work. If the temperature is reduced to 50 K for the compression back to the initial volume, the reversible compression work by the surroundings is

$$W_{com} = (0.082)(50) \ln \left(\frac{1}{10} \right) = -9.4 \, \text{Latm}$$

9.44 Latm of the 23 Latm done restores the initial volume. The remaining work, $23-9.44 = 13.56$ Latm, is available as work that can be used externally for each cycle.

In a Carnot cycle, the gas is cooled to the lower temperature using a reversible adiabatic expansion to reduce the internal energy. The gas is then compressed reversibly at the lower temperature and then compressed adiabatically and reversibly to return to the initial volume and temperature. The energy changes for the two

M.E. Starzak, *Energy and Entropy*, DOI 10.1007/978-0-387-77823-5_4,
© Springer Science+Business Media, LLC 2010

adiabatic steps are equal and opposite since they move between the same two temperatures. The net work is the difference between the isothermal, reversible expansion and isothermal, reversible compression.

4.2 The Carnot Cycle

The Carnot cycle has four steps:

(1) a reversible isothermal expansion of the gas from V_1 to V_2 at T_h;
(2) a reversible, adiabatic expansion of the gas from V_2 to V_3 with a temperature decrease from T_h to T_c;
(3) a reversible isothermal compression of the gas at T_c from V_3 to V_4; and
(4) a reversible adiabatic compression from V_4 to V_1 to restore the initial temperature T_h (Fig. 4.1)

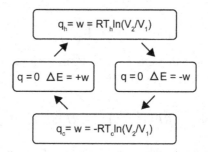

Fig. 4.1 The Carnot Cycle

The work for a reversible, isothermal expansion of 1 mol of an ideal gas at T_h from V_1 to V_2

$$w = RT_h \ln \frac{V_2}{V_1}$$

acquires its energy from an equivalent transfer of heat into the system since $\Delta E = 0$ for the ideal gas

$$\Delta E = 0 = q_{rev} - RT \ln \left(\frac{V_2}{V_1} \right)$$

$$q_{rev} = -w_{rev} = RT \ln \left(\frac{V_2}{V_1} \right)$$

The gas expands to V_3 during the adiabatic reversible expansion that decreases the temperature to T_c. $q_{rev} = 0$ and

$$\Delta E = q + w = w_{rev} = C_v (T_c - T_h)$$

The isothermal compression at T_c requires reversible work on the system and releases heat to the surroundings. This ideal engine needs heat conducting (diathermal) walls for the two isothermal changes and insulated walls during the adiabatic changes. Real engines are designed to mimic this theoretical behavior.

At T_c, the gas is compressed reversibly and isothermally from V_3 to V_4

$$w = RT_c \, \ln \frac{V_4}{V_3}$$

V_4 is selected so that an adiabatic compression returns the system to T_h and V_1.

$$\Delta E = C_V \left(T_h - T_c\right) = w$$

The temperatures and volumes for the reversible adiabatic expansion are related as

$$C_V \ln \left(\frac{T_c}{T_h}\right) = -R \ln \left(\frac{V_3}{V_2}\right)$$

while the adiabatic reversible compression relates temperatures and volumes as

$$C_V \ln \left(\frac{T_h}{T_c}\right) = -R \ln \left(\frac{V_1}{V_4}\right)$$

Since the logarithmic temperature ratio is common to both equations, the four volumes are related as ratios

$$-R \ln \left(\frac{V_3}{V_2}\right) = C_V \ln \left(\frac{T_c}{T_h}\right) = -R \ln \left(\frac{V_4}{V_1}\right)$$

and

$$-R \ln \left(\frac{V_3}{V_2}\right) = -R \ln \left(\frac{V_4}{V_1}\right)$$
$$\frac{V_3}{V_2} = \frac{V_4}{V_1}$$

or

$$\frac{V_2}{V_1} = \frac{V_3}{V_4}$$

The volume ratios are the same for the isothermal expansion and compression. The two volume ratios produce isothermal works that are opposite in sign.

The sum of the isothermal expansion and compression works is the net work. Since the volume ratios are identical for the isothermal expansion and compression, any difference in the work for these two steps depends exclusively on the temperature difference. As the difference between the temperatures in the two isothermal steps increases, the net work done during the cycle increases.

The four steps in the Carnot cycle have the following changes:

1. The isothermal reversible expansion at T_h: $\Delta E = 0$ and

$$q = -w = RT_h \ln (V_2/V_1)$$

The expansion draws heat from the surroundings and converts it to work that is stored for use in the cycle.

2. The adiabatic expansion, $q = 0$ and

$$\Delta E = w = C_v (T_c - T_h)$$

3. The isothermal compression at T_c: $\Delta E = 0$ and

$$q = -w = -RT_c \ln (V_2/V_1)$$

The piston is compressed and heat is released to the surroundings at this lower temperature.

The adiabatic compression energy and work are equal and opposite to the adiabatic expansion energy and work

$$\Delta E = w = C_v (T_h - T_c) = -C_v (T_c - T_h)$$

ΔE is the state function and its sum for the four steps of the cycle is 0. Work and heat both give non-zero sums for the four steps. Work is applied to the surroundings in the cycle

$$q_{tot} = q_1 + q_3 = RT_h \ln \left(\frac{V_2}{V_1}\right) + RT_c \ln \left(\frac{V_3}{V_4}\right) = (T_h - T_c) R \ln \left(\frac{V_2}{V_1}\right) = -w_{net}$$

The net work is proportional to the temperature difference between the hot and cold surroundings. The minus sign on net work reflects its transfer from the system to surroundings.

4.3 The Efficiency of Ideal Carnot Engines

The net work in a Carnot cycle depends on the temperature difference. For convenience and convention, net work is assigned a positive sign

$$W_{net} = (T_h - T_c) R \ln \left(\frac{V_2}{V_1}\right)$$

The factor

$$R \ln \frac{V_2}{V_1}$$

is common to net work, heat absorbed at T_h, and heat lost at T_c and cancels for ratios of these quantities.

The efficiency e, the ratio of work produced for heat absorbed at T_h, is

$$e = \frac{w_{net}}{q_h} = \frac{(T_h - T_c)\, R \ln\left(\frac{V_2}{V_1}\right)}{T_h R \ln\left(\frac{V_2}{V_1}\right)} = \frac{(T_h - T_c)}{T_h}$$

For example, when $T_h = 400$ K and $T_c = 300$ K, the efficiency or fraction of the total heat absorbed at the higher temperature that becomes net work is

$$e = \frac{T_h - T_c}{T_h} = \frac{400 - 300}{400} = 0.25$$

i.e., 25% of all the heat absorbed can be converted into work with an ideal engine working between 400 and 300 K. For $T_h = 1000$, this fraction is

$$e = \frac{1000 - 300}{1000} = 0.7$$

For $T_h = 400$ K and a temperature difference of 100 K, a heat input of 1000 J produces 250 J of net work. The remaining 750 J of the 1000 J originally absorbed by the system is released to the surroundings during isothermal compression at T_c. This work of compression is proportional to the cold bath temperature T_c,

$$q = -w = -RT_c \ln \frac{V_2}{V_1}$$

The sum of waste heat and net work equals the total heat absorbed at the hotter temperature to conserve energy.

An engine functions best when the temperature difference is largest and practical thermal engines are designed to maximize this gradient. For example, a steam engine, using superheated steam rather than steam at 100°C, converts a larger fraction of all the energy used to heat the steam into net work. Even if an engine is not reversible, its efficiency increases with an increased temperature difference. When the hot and cold temperatures are equal, the engine produces no net work.

The mean temperature of the earth is low relative to the temperature of the sun. A thermal engine might be developed to exploit heat from the sun with its hot bath temperature of 6000 K to the earth with its cold bath temperature of 300 K. This engine has a maximal efficiency of

$$e = \frac{T_h - T_c}{T_h} = \frac{6000 - 300}{6000} = 0.95$$

In principle, a device which used the sun's energy to generate net work could convert 95% of that energy directly into work. Plants, which use the sun's energy to produce stored chemical energy, do the conversion with relatively high efficiency.

The ratio of waste heat q_c to absorbed heat q_h equals the temperature ratio.

$$\frac{q_c}{q_h} = \frac{T_c}{T_h} = \frac{300}{400} = 0.75$$

75% of the heat absorbed is released as waste heat for these two temperatures.

For ratios, the heat absorbed, q_h, is proportional to T_h, the net work, w_{net}, is proportional to the temperature difference, $T_h - T_c$, and the waste heat, q_c, is proportional to T_c.

4.4 Refrigerators and Heat Pumps

A reverse Carnot cycle absorbs heat from the cold bath at T_c. Work is done on the system and converted into additional heat to raise the temperature of the system to T_h; heat is released to the surroundings at T_h by a reversible isothermal compression. The forward Carnot cycle that absorbed 8000 J of heat at 400 K to give 2000 J of net work and 6000 J of waste heat at 300 K is reversed so that 6000 J of heat is absorbed at 300 K. Then, 2000 J of work is done on the system to raise its temperature to 400 K. All this energy (8000 J) is then released to the surroundings as heat during a reversible compression at T_h.

This reverse Carnot cycle is actually an ideal, i.e., reversible, refrigerator, where heat is extracted from the interior at T_c by evaporating a fluid in coils in contact with the interior. Work by an electric motor recompresses the fluid in a region contacting the exterior and heat is released at T_h.

For the ideal refrigerator, heat is removed from the interior with a reversible isothermal expansion at T_c. A reversible adiabatic compression heats the system to T_h. A reversible isothermal compression at T_h transfers q_h to the surroundings. The work of expansion at T_c is smaller than the work of compression at T_h. The extra energy released on compression comes from the adiabatic compression.

A refrigerator's effectiveness is measured as the waste heat removed, q_c, per the net work done on the system, w_{net}

$$\frac{q_c}{w_{net}}$$

which equals the ratio of the cold temperature and the temperature difference,

$$\frac{q_c}{w_{net}} = \frac{T_c}{T_h - T_c}$$

For a refrigerator operating between $T_c = 300$ K and $T_h = 400$ K (only for illustration, not reality), the ratio is

$$q_c/w = 300/(400-300) = 3/1$$

1000 J of work by the compressor removes 3000 J of heat from the refrigerator interior. A total of 1000 + 3000 = 4000 J of heat is released to the room. The heat released to the room is determined directly by forming the ratio of q_h and the net work

$$\frac{q_h}{w_{net}} = \frac{T_h}{T_h - T_c}$$

This ideal refrigerator releases a total heat equal to

$$Q_h = q_c + w_{net}$$

4000 J is released at T_h for 1000 J of applied work.

A heat pump uses energy from the exterior plus the work to reach the interior temperature to heat the interior. Heat is extracted from the cold exterior and compressed by doing work to release both the external heat and the work of compression. A net work of 1000 J for a temperature difference of 100°C (300 K, 400 K) transfers 4000 J of heat to the house at the higher temperature. This heat pump gives four times the heating of a conventional system where work is converted directly to heat. The effectiveness of an ideal heat pump is the total amount of heating (q_h) per the net work done on the system (w),

$$\frac{q_h}{w} = \frac{T_h}{T_h - T_c}$$

Unlike the Carnot engine, where large temperature differences are used to maximize the efficiency of heat converted to usable work, a heat pump works best when the temperature differences are smaller. If the exterior temperature is 280 K, while the inside temperature is 300 K, the ratio of heat added to the house per unit work is

$$\frac{q_h}{w_{net}} = \frac{T_h}{T_h - T_c} = \frac{300}{300 - 280} = 15$$

Each joule of electrical work produces 15 J of home heat. A normal furnace would produce 1 J for each 1 J of electrical work.

To keep the temperature gradient small for effective heat pump action, the lines that collect the exterior heat are often buried below the frost line where the temperature remains relatively high and constant.

A heat pump, run in reverse, removes heat from the interior and deposits it outside. The heat pump functions as an air conditioner; heat from the house plus the work to run the engine is transferred to the exterior.

4.5 The Carnot Cycle and Entropy

Each of the four steps in the Carnot cycle involves finite work. The isothermal, reversible expansion and compression heats are not state functions but they do depend on the state variables T and V,

$$q_h = RT_h \ln(V_2/V_1)$$
$$q_c = -RT_c \ln(V_2/V_1)$$

The two reversible isothermal heat transfers differ only in their temperatures. Dividing q_h, T_h, and q_c by T_c gives equal and opposite terms in the volumes

$$\frac{q_{rev}}{T_h} = R \ln\left(\frac{V_2}{V_1}\right)$$

and

$$\frac{q_{rev}}{T_c} = -R \ln\left(\frac{V_2}{V_1}\right)$$

The sum of this ratio around the cycle is 0

$$R \ln\left(\frac{V_2}{V_1}\right) - R \ln\left(\frac{V_2}{V_1}\right) = 0$$

since the two reversible adiabatic legs with $q_{rev} = 0$ give

$$\frac{q_{rev}}{T} = 0$$

even if T is changing.

The ratio of the reversible heat to the constant temperature

$$\Delta S = \frac{q_{rev}}{T}$$

is the entropy ΔS.

The entropy measured for an isothermal volume change

$$\Delta S = R \ln \frac{V_2}{V_1}$$

depends only on initial and final volumes, not the path.

The differential change in entropy, dS, is not

$$d(q_{rev}/T)$$

Differential entropy requires a differential, reversible heat, dq, divided by constant temperature

$$dS = \frac{dq_{rev}}{T}$$

If a finite amount of heat was transferred to a system, its temperature would rise. The change in entropy is evaluated as an integral where T can be a variable

$$\Delta S = \int \frac{dq_{rev}}{T}$$

The differential dq_{rev} for an ideal gas at constant volume

$$dE = C_v dT = dq_{rev}$$

defines dS for a differential temperature change

$$dS = \frac{dq_{rev}}{T} = \frac{C_v dT}{T}$$

If the temperature of the system is changed from T_1 to T_2 in a range where the heat capacity remains constant, the entropy changes for each temperature increment dT are summed using a definite integral

$$\Delta S = \int_{T_1}^{T_2} \frac{C_v}{T} dT = C_v \ln (T) \Big|_{T_1}^{T_2} = C_v \ln \left(\frac{T_2}{T_1} \right)$$

The entropy change for a reversible adiabatic process is 0 since

$$q_{rev} = 0$$

for such a process. Even if the temperature is not constant, the entropy change is 0,

$$\Delta S = \frac{q_{rev}}{T} = \frac{0}{T} = 0$$

The reversibility condition is absolutely necessary for determining entropy since it defines the path for an entropy change. Entropy does change for an irreversible process *but it is calculated on a reversible path.* For example, the entropy can still change for an adiabatic irreversible process even though no heat is transferred.

The entropy for an irreversible process that changes a gas volume from V_1 to V_2 uses the reversible isothermal change

$$q_{rev} = RT \ln (V_2/V_1)$$

for the entropy change

$$\Delta S = \frac{q_{rev}}{T} = R \ln \left(\frac{V_2}{V_1} \right)$$

This is true even if the actual heat transferred during the irreversible process is less than the heat transferred reversibly.

4.6 The Differential Formulation of Entropy

A differential path for E using the independent variables T and V involves separate terms for changes with T and V

$$dE = \left(\frac{\partial E}{\partial T}\right)_V dT + \left(\frac{\partial E}{\partial V}\right)_T dV$$

Since S is also a state function, it follows a similar two-step differential path

$$dS = \left(\frac{\partial S}{\partial T}\right)_{dV} dT + \left(\frac{\partial S}{\partial V}\right)_T dV$$

For T and P as the independent variables,

$$dS = \left(\frac{\partial S}{\partial T}\right)_P dT + \left(\frac{\partial S}{\partial P}\right)_T dP$$

A differential expression using independent variables P and V

$$dS = \left(\frac{\partial S}{\partial V}\right)_P dV + \left(\frac{\partial S}{\partial P}\right)_V dP$$

is valid but less convenient.

The partial derivatives for the differential expressions are inferred from existing equations for the ideal gas. The entropy change for an isothermal ideal gas expansion from V_1 to V_2

$$\Delta S = R \ln \left(\frac{V_2}{V_1}\right)$$

with functional form

$$S = R \ln (V)$$

is differentiated

$$\left(\frac{\partial S}{\partial V}\right)_T = R \frac{d \ln V}{dV} = \frac{R}{V}$$

The functional form for a pressure change of an ideal gas

$$S = -R \ln (P)$$

gives

$$\left(\frac{\partial S}{\partial P}\right)_T = -R \left(\frac{\partial \ln (P)}{\partial P}\right)_T = \frac{-R}{P}$$

A non-ideal gas with equation of state

$$P(V - b) = RT$$

has

$$\Delta E = 0 \text{ and}$$

$$P = \frac{RT}{V - b}$$

Then

$$ds = \frac{dq_{rev}}{T} = \frac{PdV}{T}$$

$$= \frac{\frac{RT}{V-b}dV}{T} = \frac{R}{V - b}dV$$

The partial differential expression for S versus T at constant V is

$$dS = \left(\frac{\partial S}{\partial T}\right)_V dT = \frac{C_V}{T} dT$$

so

$$\left(\frac{\partial S}{\partial T}\right)_V = \frac{C_V}{T}$$

For constant C_V, the entropy change for a temperature change from T_1 to T_2 is

$$\Delta S = \int dS = \int_{T_1}^{T_2} \frac{C_V}{T} dT$$

Each differential increment of heat was transferred reversibly at a constant temperature but the sum of all these incremental reversible transfers gives a net macroscopic entropy change.

The entropy for a transition between (T_1, V_1, P_1) and (T_2, V_2, P_2)

$$dS = \left(\frac{\partial S}{\partial T}\right)_V dT + \left(\frac{\partial S}{\partial V}\right)_T dV$$

is

$$dS = \frac{C_v}{T}dT + \frac{R}{V}dV$$

for an ideal gas. Integrating

$$\Delta S = C_v \int_{T_1}^{T_2} \frac{dT}{T} + R \int_{V_1}^{V_2} \frac{dV}{V}$$

$$= C_v \ln\left(\frac{T_2}{T_1}\right) + R \ln\left(\frac{V_2}{V_1}\right)$$

A new equally valid path is possible with T and P

$$dS = \left(\frac{\partial S}{\partial T}\right)_P dT + \left(\frac{\partial S}{\partial P}\right)_T dP$$

For constant P,

$$dq_{p,\mathrm{rev}} = dH = C_p dT$$

and

$$dS = \frac{dq_{p,\mathrm{rev}}}{T} = \frac{C_p}{T}dT$$

Since

$$dS = \left(\frac{\partial S}{\partial T}\right)_P dT$$

$$\left(\frac{\partial S}{\partial T}\right)_P = \frac{C_p}{T}$$

The partial derivative for the change of entropy with pressure at constant temperature for an ideal gas

$$\left(\frac{\partial S}{\partial P}\right)_T = -\frac{R}{P}$$

completes the T, P path

$$dS = \frac{C_p}{T}dT - \frac{R}{P}dP$$

4.7 Entropy Paths

The differential reversible paths (T, V and T, P) must give the same entropy change because each path takes the system from one state, characterized by its state variables T, V and P, to a final state. For example, an ideal monatomic gas in a state with $(P_1, V_1, T_1) = (1 \text{ atm}, 10 \text{ L}, 300 \text{ K})$ changes to a state with final values (4 atm, 5 L, 600 K). The entropy is calculated by either of two distinct paths. For path 1, P is constant at 1 atm while T rises from 300 to 600 K. Then, P changes from 1 to 4 atm while T is constant at 600 K. The third parameter, the volume, changes to satisfy the ideal gas law, and is not an independent variable.

C_p for an ideal monatomic gas

$$C_P = C_V + R = \frac{3R}{2} + R = 2.5R$$

is used for the (T, P) path

$$dS = \frac{\frac{5R}{2}dT}{T} - \frac{R}{P}dP$$

Integrating from initial to final temperature and pressure gives

$$\Delta S = C_p \ln\left(\frac{600}{300}\right) - R \ln\left(\frac{4}{1}\right) = (3R/2)\ln(2) - R\ln(2) = (R/2)\ln(2)$$

For the second path, V first decreases from 10 to 5 L at 300 K. The third variable, the pressure, increases to 2 atm. T increases at a constant 5 L

$$\Delta S = R\ln\frac{V_f}{V_i} + C_V \ln\frac{T_f}{T_i}$$
$$= R\ln(5/10) + (3R/2)\ln(600/300)$$
$$= (R/2)\ln(2)$$

The values for each leg of the two paths are different but the entropy change is exactly the same for each total path between the two states.

4.8 Entropy Changes of the Surroundings

The isothermal expansion of an ideal gas involves the reversible transfer of heat, q_{rev}, from the surroundings to compensate the reversible work done by the system at constant internal energy, $\Delta E = 0$. Since energy is conserved, the surroundings lose q_{rev} as determined from the calculations for the system. Since system and surroundings have the same T for this transfer, the entropy change of the surroundings is determined using this reversible heat

$$q_{rev,sur} = -q_{rev,sys}$$

to give

$$\Delta S_{sys} = -\Delta S_{sur}$$

The entropy change of the universe is 0 for reversible changes:

$$\Delta S_{univ} = \Delta S_{sys} + \Delta S_{sur}$$
$$= 0$$

A reversible adiabatic expansion has $q_{rev} = 0$ and $\Delta S = 0$. An irreversible adiabatic expansion has $q = 0$, but this does not mean the entropy change is 0. The actual entropy change is determined from the initial and final system variables.

For a reversible adiabatic expansion, the entropy changes of the system and surroundings are both 0,

$$\Delta S_{sys} = \frac{q_{rev}}{T} = \frac{0}{T} = 0$$

The entropy change of the universe, the sum of these two entropies, is also 0.

The system entropy change for an adiabatic irreversible expansion is determined from changes in the state variables (the reversible path)

$$\Delta S = C_v \ln\left(\frac{T_2}{T_1}\right) + R \ln\left(\frac{V_2}{V_1}\right)$$

since volume and temperature changes are known. The expansion uses the system's internal energy so volume increases, as temperature decreases for the adiabatic expansion.

No heat is transferred in this irreversible expansion. $q_{sur} = 0$. The entropy for the surroundings is calculated from the actual heat ($q=0$) transferred to the surroundings; this transfer is assumed reversible

$$q_{sur,irr} = 0$$

and

$$\Delta S_{sur} = \frac{q_{rev}}{T} = \frac{0}{T} = 0$$

The volume entropy change of the system increases more than the temperature entropy change decreases so their sum is a net positive. The entropy of the universe is positive. In general, the entropy of the universe increases for any spontaneous (non-reversible) process. Entropy is not conserved; it is produced during irreversible changes.

The first two laws of thermodynamics are often expressed concisely as (1) energy is conserved and (2) the entropy of the universe increases. The last statement reflects the fact that very few changes are reversible. Almost all proceed irreversibly with the production of entropy.

4.9 Reversible Paths for Entropy in Irreversible Changes

The Joule experiment expands gas irreversibly into an evacuated bulb. The gas does not spontaneously return to the initial volume in the left bulb. The entropy of the system, however, is calculated by a reversible path, not the actual path. One mole of ideal gas in a 1 L bulb expands into a second evacuated 1 L bulb so that the gas occupies a final volume of 2 L with no temperature change for the ideal gas.

The entropy is calculated via the reversible path where only volume changes

$$\Delta S = R \ln \left(\frac{V_2}{V_1} \right) = R \ln \frac{2}{1}$$

The system is insulated so $q_{sur} = 0$. The reversible "path" for the surroundings is

$$\Delta S_{sys} = q_{rev}/T = 0$$

and

$$\Delta S_{univ} = \Delta S_{sys} + \Delta S_{sur} = R \ln \left(\frac{2}{1} \right) + 0 = R \ln (2)$$

The entropy of the universe increases for the irreversible Joule expansion.

There is a strong correlation between the generation of entropy and spontaneity. A gas always expands into a vacuum; heat always flows from the hotter to the colder temperature. These are spontaneous processes that generate entropy.

Heat always flows spontaneously from the hotter to the colder. For a reversible Carnot cycle, the entropy change is 0 and the net work is maximal.

4.10 Non-equilibrium Phase Transitions

Phase transitions occur at a constant temperature. For example, liquid water becomes vapor at 100°C (373 K) and 1 atm and freezes to ice at 273 K and 1 atm pressure. During the transition, heat is transferred. If the phase transition occurs at constant pressure,

$$\Delta H_{\text{phase transition}} = q_P$$

The enthalpy of vaporization, ΔH_{vap}, is positive since heat is added to the system to vaporize a liquid. Since energy is conserved, this energy is removed from the surroundings as heat

$$\Delta H_{sur} = -\Delta H_{vap}$$

In the reversible limit, the phase transition occurs when the temperatures of both system and surroundings are equal at the transition temperature.

For a vaporizing system at the boiling point,

$$\Delta S_{vap} = \frac{q_{rev}}{T_{bp}} = \frac{\Delta H_{vap}}{T_{bp}}$$

while

$$\Delta S_{sur} = -\frac{\Delta H_{vap}}{T_{bp}}$$

and

$$\Delta S_{univ} = \Delta S_{sys} + \Delta S_{sur} = \frac{\Delta_{vap}}{T_{bp}} + \frac{-\Delta H_{vap}}{T_{bp}} = 0$$

A reversible phase transition transfers entropy into the system without generating new entropy at a constant temperature. A phase transition when system and surroundings temperatures are different generates entropy.

A substance freezes reversibly at 300 K, releasing 3000 J mol^{-1} with an entropy decrease of

$$\Delta S_{fus,sys} = \frac{-3000}{300} = -10 \, J \, K^{-1} mol^{-1}$$

At 300 K, the surroundings' entropy increases the same amount

$$\Delta S_{fus,sur} = +10 \, J \, K^{-1} \, mol^{-1}$$

This liquid is supercooled to 290 K where it can freeze spontaneously. The enthalpy and entropy of fusion at this temperature are determined for a reversible path (Fig. 4.2).

(1) The supercooled liquid is heated to 300 K.
(2) A reversible phase transition occurs at 300 K
(3) The solid is cooled to 290 K.

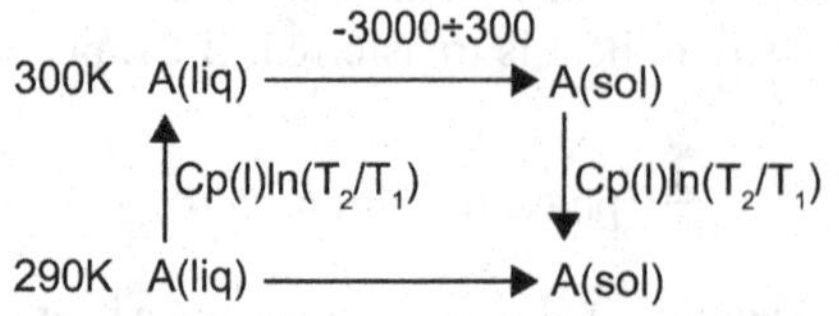

Fig. 4.2 A reversible path for the transition from supercooled liquid to solid

The enthalpy change for fusion at 290 is

$$\Delta H = C_p(\text{liq})(300 - 290) - 3000 + C_p(\text{sol})(290 - 300)$$

If $C_p(1) = 6$ and $C_p(s) = 4$,

$$\Delta H = -3000 + 6(10) + 4(-10) = -2980$$

The enthalpy changes only slightly with temperature.
The entropy on the reversible path is

$$C_p(\text{liq})\ln(300/290) - 3000/300 + C_p(\text{sol})\ln(290/300)$$

The entropy change of the system is

$$6(0.034) - 10 + 4(-0.034) = -5 + 2(0.034) = -9.9$$

The entropy change of the system is smaller than the reversible change but the entropy of the universe must increase. The entropy of the surroundings is found from the enthalpy released at 290 K.

$$\Delta H = 6(300 - 290) + (-1500) + 4(290 - 300) = 60 - 3000 - 40 = -2980$$

The entropy of the surroundings is found by assuming this heat is added to the surroundings reversibly

$$\Delta S_{\text{sur}} = +2980/290 = +10.3 \, \text{J mol}^{-1}\text{K}^{-1}$$

The entropy of the universe

$$-9.9 + 10.3 = 0.4 \, \text{J mol}^{-1} \, \text{K}^{-1}$$

increases for this irreversible process.

Problems

4.1 An ideal monatomic gas expands from a volume of 2 L when a pressure $P_{\text{ext}} = P_{\text{int}} = 5$ atm is applied. The gas flows through a porous material to a second compartment with $P_{\text{ext}} = P_{\text{int}} = 2$ atm. The volume of this compartment increases from 0 to 5 L. The temperature is a constant 300 K.
  Determine ΔS_{sys}
4.2 One mole of an ideal gas with $C_V = 1.5R = 12.5 \, \text{J mol}^{-1} \, \text{K}^{-1}$ and

$$\left(\frac{\partial E}{\partial V}\right)_T = 20 \, \text{J L}^{-1}$$

expands against a constant external pressure $P_{ext} = 0.2$ atm from 1 to 11 L (1 Latm $=100$ J). During the change, the temperature decreases from 350 to 300 K.

a. Determine the entropy change of the system.
b. Determine the entropy change of the surroundings at constant 300 K.

4.3 The adiabatic expansion in a Carnot cycle lowers the temperature of 1 mol of an ideal monatomic gas ($C_V = 12.5$ J K^{-1} mol^{-1}) from 350 to 300 K.

a. Determine the work done during this reversible adiabatic expansion step.
b. Determine the entropy change of the system during the reversible adiabatic expansion step.

4.4 The volume of 1 mol of an ideal monatomic gas initially at 2 atm, 300 K, and 12.2 L is doubled by (a) reversible isothermal expansion and (b) reversible adiabatic expansion. Calculate the entropies of the system and surroundings for each case.

4.5. An ideal heat pump operating between an outside temperature of 270 K and an inside temperature of 300 K deposits 1200 J of heat in the house. Determine how much heat would be deposited if the same amount of heat were drawn from the outside but the engine was only 50% efficient, i.e., it had to do twice as much work as the ideal machine to extract the same amount of heat from the outside.

4.6 A mole of solid material at 400 K with a heat capacity of 25 J K^{-1} mol^{-1} at all temperatures is dropped into a large water bath with a temperature of 300 K. The heat from the metal is then lost irreversibly to the water although the water bath is so large that the temperature of the bath remains effectively constant as it receives the heat from the metal.

a. Determine the entropy change for the block. The volume change of the block on cooling is very small and can be ignored.
b. Determine the entropy change for the constant temperature water bath.
c. Determine the entropy change for the universe.

4.7 The entropy of gas is defined by

$$\left(\frac{\partial S}{\partial V}\right)_T = \frac{R}{V-b} \qquad \left(\frac{\partial S}{\partial T}\right)_V = \frac{C_v}{T}$$

Find the entropy change when the system changes from (V_1, T_1) to (V_2, T_2).

4.8 One mole of an ideal gas with $C_p = 7R/2 = 30$ J K^{-1} mol^{-1} and $(\partial E/\partial V)_T = 0$ absorbs 1100 J of heat during an irreversible expansion. The temperature changes from 200 to 400 K (the surroundings temperature), while the pressure is constant at 8.2 atm. The initial volume is 2 L.

Calculate the entropy change for system and surroundings.

4.9 One mole of an ideal diatomic gas expands in two steps: an adiabatic reversible expansion followed by an isothermal reversible expansion. The initial temperature and pressure of 300 K and 1 atm reach a final temperature and pressure of 200 K and 0.25 atm, respectively.

a. Determine the entropy change for the isothermal reversible expansion alone.
b. Determine the entropy change for the surroundings.

Chapter 5
The Nature of Entropy

5.1 The Nature of Entropy

Entropy is different from conserved energies and system variables like T and V that can be measured in the laboratory. However, the product TS has units of energy which suggests that entropy produced in an irreversible process also produces energy. Both the entropy S and the product TS are related to the randomness of the system. The connection between randomness and both entropy and energy is one of the more interesting concepts in thermodynamics. Its physical meaning can be approached probabilistically

For isothermal, reversible expansions, the entropy depends only on the change of volume for an ideal gas:

$$\Delta S = R \ln \left(\frac{V_2}{V_1} \right)$$

This same entropy change could occur at any temperature. Entropy in this case only depends on the change of the system volume.

During a vaporization, heat is absorbed as liquid becomes gas but the temperature of the two-phase system is constant. Energy is added but does not appear as an increase in the kinetic energy of the system molecules. The entropy increase during a vaporization at constant temperature

$$\Delta S_{vap} = \frac{\Delta H_{vap}}{T}$$

has to be associated with the large volume change in the liquid to gas transition. The energy is added as heat to increase the entropy but not the kinetic energy of the system.

5.2 Trouton's Rule

Although enthalpies of vaporization and boiling point temperature can vary significantly for different liquids, the entropy of vaporization is roughly consistent.

M.E. Starzak, *Energy and Entropy*, DOI 10.1007/978-0-387-77823-5_5,
© Springer Science+Business Media, LLC 2010

Benzene, which boils at 353.25 K with $\Delta H_{vap} = +30{,}800$ J mol^{-1}, gives an entropy increase of

$$\Delta S = \frac{\Delta H_{vap}}{T_{bp}} = \frac{+30{,}800}{353.25} = 87.2\,\mathrm{J\,mol^{-1}K^{-1}}$$

Liquid nitrogen which boils at 77.4 K with $\Delta H_{vap} = 5818$ J mol^{-1} has

$$\Delta S = \frac{5818}{77.4} = 75.2\,\mathrm{J\,K^{-1}\,mol^{-1}}$$

Trouton's rule states that the entropy of vaporization is approximately 85 J mol^{-1} K^{-1}. Water and NH_3 with considerable hydrogen bonding (and structure) in the liquid phase, i.e., more liquid order, have higher entropies of vaporization (ca. 110 J K^{-1} mol^{-1}).

Trouton's rule permits estimates of the enthalpies of vaporization if the boiling point is known. The enthalpy of vaporization of carbon tetrachloride with a boiling point of 349.9 K is approximately

$$\Delta H = T_{bp}85\,\mathrm{J\,K^{-1}\,mol^{-1}} = 29{,}665\,\mathrm{J\,mol^{-1}}$$

which compares very favorably with the experimental vaporization enthalpy of 30,000 J mol^{-1}.

A gas occupies a much larger volume than a liquid. Entropy is a measure of randomness. Each molecule is located in many more locations as a gas. Ammonia and water have this locational entropy increase plus the entropy increase when the hydrogen bonded liquid structure is destroyed on vaporization.

If the entropy increase on vaporization is due to increased vapor volume, the entropy can also be estimated using the volume change on vaporization:

$$\Delta S_{vap} = R \ln \left(\frac{V_{vap}}{V_{liq}} \right)$$

A mole of liquid might occupy about 2 mL, while the same gas at 25°C occupies 30 L (30,000 mL):

$$\Delta S = R \ln \left(\frac{V_2}{V_1} \right) = 8.31\,\ln \left(\frac{30{,}000}{2} \right) = 80\,\mathrm{J\,K^{-1}\,mol^{-1}}$$

The volume approach gives a value comparable to that of Trouton's rule (85 J K^{-1}mol^{-1}). A more accurate calculation includes molecular interactions and the degree of randomness in the liquid.

Solids and liquids are more dense than gases and the molecules can interact strongly. A phase transition between a solid and a liquid or a solid and a solid phase involves a change in crystal structure and interactions. Entropy changes for these phase transitions vary considerably and there is no equivalent of Trouton's rule for liquid–solid and solid–solid phase transitions.

5.3 Volume Changes and Randomness

The entropy change for an isothermal ideal gas depends only on the change in the gas volume. The ratio of volumes for the entropy change

$$\Delta S = R \ln \left(\frac{V_2}{V_1} \right)$$

is equivalent to the difference

$$\Delta S = S_2 - S_1 = R \ln (V_2) - R \ln (V_1)$$

The entropy of a state is directly proportional to the natural logarithm of its volume:

$$S = R \ln (V)$$

This entropy can be divided by Avogadro's number to give an entropy per molecule:

$$S' = S/N_a = R/N_a \ln (V) = k \ln (V)$$

k with units of J molecule^{-1} K^{-1} is Boltzmann's constant. This individual molecule entropy still has the logarithmic volume dependence. A single molecule can increase its entropy if it is free to move (at any speed) in a larger volume. The molecule is harder to locate in the larger volume no matter how fast it moves.

5.4 States

One mole of an ideal gas that expands isothermally from 1 to 2 L has an entropy change

$$\Delta S = R \ln \left(\frac{2}{1} \right)$$

For one molecule,

$$\Delta S = k \ln (2/1)$$

Each molecule that was free to move anywhere in 1 L is now free to move anywhere in 2 L.

Each volume is partitioned into a set of cubes of equal volume. A random molecule "visits" each of these cubes with equal probability. Ideally, the cube volume is fine grained, i.e., each microvolume is molecular size. However, entropy differences are determined from volume ratios. As long as 2 L has twice as many cubes as 1 L, the entropy change is known.

Each of the volume cubes is called a state of the system. The volume doubling from 1 to 2 L is equivalent to doubling the number of location states (Ω)

$$\Delta S = R \ln \left(\frac{2\Omega}{\Omega} \right)$$

The entropy change for just one molecule is

$$\Delta S = k \ln \left(\frac{2\Omega}{\Omega} \right)$$

The cubes are equally accessible to this single molecule.

Intuitively, two molecules might be expected to occupy twice the number of states. For a volume cube of 1 L, 1 L is one state and 2 L is two states. Two molecules produce four (2^2) distinct states: (1) both particles in the left liter; (2) both particles in the right liter; (3) particle 1 in the left and particle 2 in the right; and (4) particle 2 in the left and particle 1 in the right.

Three molecules expanding from 1 to 2 L increase from 1 state (all in the left "cube") to $2^3 = 8$ states; each molecule has twice the possible locations. An Avogadro's number of molecules increases from 1 to

$$2^{N_a}$$

states. A volume increase from 1 to 3 L for N_a particles gives

$$\left(\frac{3}{1} \right)^N$$

as the state ratio.

A difference in entropy is related to a ratio of states.

Boltzmann defined an absolute entropy for Ω states as

$$S = k \ln (\Omega)$$

with $k_b = 1.38 \times 10^{-23}$ J K^{-1} molecule^{-1}. While location state volume is arbitrary, definite states can be allotted to some systems. This important equation appears on Boltzmann's tombstone in Vienna. This might have been his only equation that fitted.

5.5 States and Probability

In Boltzmann's formulation, all states are accessible to the molecule. Each state is equally probable. If the number of states is known, the probability of being in a specific state is the inverse of number of states. A system with 100 states has a probability of $1/100 = 0.01$ that a molecule is found in a specific state. Doubling the size and the states of the system to 200 decreases the probability to $1/200 = 0.005$. The probability of finding the particle in a specific state is halved. The increased randomness of the system is associated with a decreased probability.

A two-particle system that goes from 1 to 2 states has a probability

$$1/2^2 = (1/2)^2 = 1/4$$

One quarter of the time, both particles will return to their initial state.

The probability of returning to the original state decreases dramatically with increasing particles (N). The probability that 10 particles expanded from 1 to 2 L will return to their original 1 L is

$$1/2^{10} = 1/1024 = 0.00098$$

An Avogadro's number of particles has a return probability

$$1/2^N$$

that is vanishingly small. This low probability is responsible for the irreversible nature of the expansion from 1 to 2 L. The probability that all the particles, each undergoing its own random motion, all collect in the left 1 L is vanishingly small. Once the gas expands, it stays expanded.

The gas, on an average, cycles through all possible states before it returns to the original state. This is a Poincare cycle. The system with ten particles and 1024 states passes through 1024 states, i.e., all the states, before it can return to the initial state. If the average time T to pass from one state to the next is known, the Poincare recurrence time is just

$$T\Omega$$

In Boltzmann's formulation, all states are equally probable. However, only one of these equally probable states places all the particles on the left. This is a $(L, R) = (N, 0)$ distribution. The 50–50 ($N/2$, $N/2$) distribution appears much more often since so many distinct states give this distribution.

5.6 Entropy and Temperature

An isothermal increase in the volume increases the number of states and this leads to an increase in the system entropy. The kinetic energy plays no role at all. However, the entropy also increases with temperature

$$\Delta S = C_v \ln \left(\frac{T_2}{T_1} \right)$$

at constant volume. If kinetic energy is not involved, what produces this change in entropy?

The kinetic energies determined by the heat capacities are averages. The particles actually have a distribution of energies. For translation, some particles move faster, while others move slower, i.e., a velocity distribution or energy distribution. The range of velocities and energies increases at higher temperatures. If possible molecular velocities define states, the higher temperature system has more such states. The entropy change

$$\Delta S = C_v \ln (T_2/T_1)$$

reflects this increase in velocity states.

A system at higher temperature does net work as it drops to the lower temperature. The loss of high energy states releases some of the randomness as work energy.

Entropy is a system variable like temperature, pressure, or volume. In the reversible limit, the first law becomes

$$dE = dq_{rev} + dw_{rev} = TdS - PdV$$

All external variables are replaced by state (system) variables.

5.7 The Entropy of Mixing

The mixing of two liquids, e.g., coffee and cream, or two different gases is generally a spontaneous and irreversible process. Once mixed, coffee and cream do not separate. The mixed system is more disordered or random and entropy is larger. Mixing is an irreversible process; the mixed materials never spontaneously unmix. However, the materials must be mixed reversibly to calculate the entropy of mixing.

The irreversible mixing of n_A moles of ideal gas A in a volume V_A and n_B moles of ideal gas B in volume V_B at the same constant temperature and pressure occurs by joining two cylinders with these gases. The total volume

$$V_t = V_A + V_B$$

is proportional to $n_t = n_A + n_B$. Temperature and pressure remain constant.

If gas A expands from V_A to V_t,

$$V_t = V_A + V_B$$

the entropy change is

$$\Delta S_A = n_A R \ln \left(\frac{V_t}{V_A} \right)$$

If B expands isothermally from V_B to V_t, the entropy change is

$$\Delta S = n_B R \ln \left(\frac{V_t}{V_A} \right)$$

These entropy changes do occur for the two gases when they each expand during mixing. However, the calculations do not include an actual entropy change for mixing. Mixing requires an additional reversible step where the two gases can mix or separate.

A reversible path to calculate the entropy need not correspond to the path actually followed by the system from its initial to its final state. In fact, the path need not be an experimentally viable path. A hypothetical, reversible path gives the exact entropy change. For the entropy of mixing, Planck proposed a gedanken (thought) experiment for reversible mixing (Fig. 5.1). Gases A and B expand to V_t in separate cylinders. Planck postulated special caps for one end of each expanded cylinder. The cylinder containing gas A is capped with an end plate which, while impermeable to A, is completely permeable to B. A cannot escape from this cylinder but B can enter through the cap. The cylinder containing B had an end cap which is permeable to A. For a reversible mixing, the cylinders, each of volume V_t, are arranged with their special end caps facing each other. The cylinders are pushed together to form one cylinder with total final volume V_t. A molecules pass easily through the cap of cylinder B to reach the common region, while B molecules pass easily through the cap of cylinder A to reach this same common region. The A and B separate reversibly as the cylinders are pulled apart.

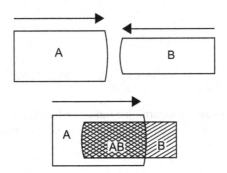

Fig. 5.1 Planck reversible mixing

The total entropy for mixing is the sum of the three reversible steps. A and B are expanded to the volume V_t in their separate cylinders. The reversible mixing step involves no change in internal energy (temperature constant) and no work ($P_{ext} = 0$). Since

$$\Delta E = 0 = q + w = q = q_{rev}$$

The entropy change for this third (mixing) step is 0! The total entropy of mixing is

$$\Delta S = n_A R \ln \left(\frac{V_t}{V_A} \right) + n_B R \ln \left(\frac{V_t}{V_B} \right) + 0$$

T and P are constant for this mixing. Although the individual pressures decrease on expansion, the total pressure is still P

$$P_{A,f} = P_{A,i} \frac{V_A}{V_t}$$

$$P_{B,f} = P_{B,i} \frac{V_B}{V_t}$$

$$P_{A,f} + P_{B,f} = P \left(\frac{V_A}{V_t} + \frac{V_B}{V_t} \right) = P \frac{V_A + V_B}{V_t} = P$$

Since the temperature and total pressure are constant

$$n_t = n_A + n_B = \frac{P}{RT} (V_A + V_B) = \frac{P}{RT} V_t$$

the ratio $\{V_t\}$ over $\{V_A\}$ is

$$\frac{V_t}{V_A} = \frac{n_t \frac{RT}{P}}{n_A \frac{RT}{P}} = \frac{n_t}{n_A}$$

The B ratio is

$$\frac{V_t}{V_B} = \frac{n_t}{n_B}$$

Substituting these mole fractions gives

$$\Delta S_{\text{mix}} = n_A R \ln \left(\frac{n_t}{n_A} \right) + n_B R \ln \left(\frac{n_t}{n_B} \right) = -R [n_A \ln (X_A) + n_B \ln (X_B)]$$

The sign changes when the logarithmic ratios are inverted.

This total entropy of mixing for $n_t = n_A + n_B$ moles is converted to an entropy per total moles

$$\Delta \bar{S}_{\text{mix}} = \frac{\Delta S}{n_t} = -R \left(\frac{n_A}{n_t} \right) \ln (X_A) + \frac{n_B}{n_t} \ln (X_B) = -R (X_A \ln X_A + X_B \ln X_B)$$

The entropy of a mixture of N species with mole fractions, X_i, is

$$\Delta \bar{S}_{\text{mix}} = -R \sum_{i=1}^{i=N} X_i \ln X_i$$

Mixing gases at different pressures or temperatures is done by first determining the entropies required to bring the temperature and pressures to their values for mixing. If one mole of a monatomic gas A at a pressure of one atmosphere and a temperature of 300 K is mixed with one mole of a second monatomic gas B at 1 atm pressure and 400 K, the final temperature of the mixture will be 350 K at 1 atm total pressure because both gas heat capacities are equal. The entropy of this mixture is found in two steps:

(1) the entropy required to change the temperatures to the final temperature of the mixture and
(2) the entropy of mixing at 350 K and 1 atm.

The entropy change to bring both gases to 350 K is

$$\Delta S = (1)\, C_{p(A)} \ln \frac{350}{300} + (1)\, C_p \ln \frac{350}{400} = \frac{5}{2} R \left(\ln \frac{7}{6} + \ln \frac{7}{8} \right) = 0.051R$$

for the two moles of gas. The entropy of mixing for the two moles of gas is

$$\Delta S_{\text{mix}} = 2 \Delta \bar{S}_{\text{mix}} = -2R\, (0.5 \ln 0.5 + 0.5 \ln 0.5) = 1.39R$$

The entropy change for mixing dominates because mixing is a more spontaneous process. The total entropy is

$$\Delta S_t = \Delta S_T + \Delta S_{\text{mix}} = 0.051R + 1.39R = 1.44R$$

5.8 Entropy of Mixing for Microscopic States

The molar entropy of mixing

$$\Delta S = -R \sum_i X_i \ln X_i$$

is converted to an entropy of mixing for a single molecule by dividing both sides by Avogadro's number

$$\Delta S = -k_b \sum X_i \ln X_i$$

How is it possible to have an entropy of mixing for one molecule? The molecule is either A or B. To resolve this anomaly, the mole fraction for A for a large number of particles is equivalent to the probability that a molecule A is selected if one particle is pulled at random from the mixture. For one particle that is either A or B, $X_A = p_A$ and $X_B = p_B$. The probabilities indicate something unknown about a single particle. If $p_A = p_B = 0.5$

$$\Delta S = -k_b \{0.5 \ln (0.5) + 0.5 \ln (0.5)\} = -k_b \ln (0.5) = +k_b\, (0.69)$$

is the largest possible for a two species mixture. Since the two are equally probable, the choice is most random. Neither guess is favored. If the probabilities are

$$p_A = 0.8 \text{ and } p_B = 0.2$$

a gambler who guessed that A would be picked has $0.8/0.2 = 4/1$ odds. This order is reflected in a lower entropy

$$\Delta S = -k \left(0.8 \ln (0.8) + 0.2 \ln (0.2)\right) = 0.5\, k$$

The Boltzmann equation

$$S = k \ln (\Omega)$$

is generated from the entropy of mixing when Ω identical states are mixed. The probability of each state is $1/\Omega$ and the entropy of mixing

$$S = -k \sum_{1}^{\Omega} \frac{1}{\Omega} \ln \left(\frac{1}{\Omega}\right) = -k\Omega \left(1/\Omega\right) \ln \left(\frac{1}{\Omega}\right) = +k \ln (\Omega)$$

5.9 The Third Law of Thermodynamics

The entropy of mixing is the change from unmixed to mixed. The unmixed state is assigned a probability of 1 since A and B are located exactly in their respective containers. The entropy difference on mixing is

$$\Delta S = -k \left[(X_A \ln X_A - 1 \ln 1) + (X_B \ln X_B - 1 \ln 1\right]$$
$$\Delta s = s_f - s_i = s_f - 0 = s_{mix}$$

Since $\ln(1) = 0$.

The entropy of mixing is relative to a starting entropy of 0 even though each gas does have an entropy.

Boltzmann's equation

$$S = k \ln (\Omega)$$

is not a difference equation. The reference state consists of one state and an entropy of 0.

A crystalline solid of atoms confines each atom to a single volume element. The volume for each atom expands with increasing temperature as the whole crystal expands since each atom moves within its own cavity. As the temperature approaches absolute zero, the kinetic motions decrease locking each atom in a stationary position in classical models of a crystal.

At absolute zero, the kinetic energy of each classical atom is 0; the atoms are motionless in their cavities. At absolute zero, the entire crystal is one state. Each atom can be located exactly. The absolute entropy of an atom in the crystal at 0 K is

$$S = k_B \ln (\Omega) = k_B \ln (1) = 0$$

The third law of thermodynamics states that the entropy of a perfect atomic crystal at absolute zero is 0.

As the volume or temperature of the crystal increases, the corresponding entropy change is actually an absolute entropy since its reference entropy is 0 at $t = 0$ K

$$\Delta S = S_f - S_i \ (T = 0\,K) = S\,(T,V) - 0$$

The absolute entropy of the elements in their most stable states at 25°C and 1 atm is determined by summing all the entropy changes from 0 K. Tables of absolute entropies are listed at 25°C and 1 atm so that they can be used in conjunction with internal energy and enthalpy changes.

5.10 Absolute Entropy Calculations

The Debye theory for heat capacities based on quantized energies states that, near zero, heat capacity follows a "T-cubed" law

$$C_p = aT^3$$

where a is a constant for each crystal.

The absolute entropy at any temperature and pressure is the sum of entropy change from $T = 0$ K. Near zero, the T-cubed law is used

$$S\,(T) = \int_0^T \frac{C_p}{T}dT = \int_0^T \frac{aT^3}{T}dT = \frac{aT^3}{3}\bigg|_0^T = \frac{aT^3}{3}$$

Further integration from T to the melting point (or phase transition temperature) is approximated with a constant heat capacity:

$$\Delta S = \int_T^{T_{mp}} \frac{C_p}{T}dT$$

At the melting point, the entropy of the phase transition from the solid to the liquid is

$$\Delta S = \frac{\Delta H_{mp}}{T_{mp}}$$

If the material is liquid at 25°C and 1 atm, the liquid entropy change is added:

$$\Delta S_{liq} = \int_{T_{mp}}^{298} \frac{C_p\,(liq)}{T}dT$$

If the material is gas at the final T and P, the entropy for the liquid–vapor phase transition and changes in the pressure and temperature of the gas are added for a total absolute entropy.

5.11 Residual Entropy

A perfect crystal at a temperature of 0 K has zero entropy because each atom or molecule is confined to its own volume cell. The entropy is larger than 0 for imperfect crystals. However, some perfect crystals can have a finite entropy at absolute zero.

The entropy at 0 K is calculated using either thermodynamics or statistical thermodynamics. Although the two approaches should give identical results, a difference, the residual entropy, is observed for certain molecules. For example, the entropies for the CO molecule differ by $R\ln(2)$.

C and O atoms in CO are roughly the same size. A CO molecule can fit into its lattice site as either CO or OC in the perfect crystal. The two orientations become two states at 0 K. Instead of a single perfect crystal, $\Omega=1$, each crystal site has two equally probable orientations. For N such sites (or molecules), there are

$$2^N$$

distinct perfect crystals for CO. The absolute entropy at 0 K is

$$S = k_b \ln \left(\Omega^N\right) = R\ln\left(\Omega\right) = R\ln\left(2\right)$$

Symmetrical linear molecules like O_2 and CO_2 have no residual entropy. Quantum mechanics states that two-like atoms are indistinguishable, i.e., the oxygen atoms can not be labeled. Without labels, two orientations are indistinguishable.

The residual entropy for crystalline ice has 3/2 states

$$S = R\ln\left(3/2\right)$$

Linus Pauling explained the factor by noting water is a tetrahedron with two protons and two unpaired electrons that has six different spatial orientations in space

$$4!/2!2! = 6$$

The number of orientational states is reduced because the only allowed orientations have the water's protons face electron pairs on adjacent molecules (2/4 possibilities) and its electron pairs must face protons on adjacent molecules (2/4). This reduces the number of possible residual states as

$$6\left(2/4\right)\left(2/4\right) = 6/4$$

and the residual entropy is

$$S = k_b \ln \left(\frac{6}{4}\right)^{N_a} = R\ln\left(1.5\right)$$

5.12 The Gibbs Paradox

A 50–50 mixture of A and B with $X_A = X_B - 0.5$ has entropy of mixing

$$\Delta S = R \ln (2)$$

For example, 0.5 mol of O_2 mixed with 0.5 mol of N_2 gives X_{O2} and $X_{N2} = 0.5$ and an entropy of mixing

$$S = -R (0.5 \ln (0.5) + 0.5 \ln (0.5)) = R \ln (2)$$

A cylinder containing 0.5 mol of $^{16}O^{16}O$ can be mixed with a second cylinder containing 0.5 mol of isotopic oxygen $^{18}O^{18}O$ to produce the same entropy of mixing,

$$\Delta \bar{S}_{mix} = -R[0.5 \ln 0.5 + 0.5 \ln 0.5] = R \ln 2$$

If 0.5 mol of $^{16}O^{16}O$ is mixed with 0.5 mol of $^{16}O^{16}O$, the predicted entropy of mixing is

$$\Delta \bar{S}_{mix} = R \ln 2$$

However, if these molecules are indistinguishable, the final "mixture" looks exactly like the original, unmixed system with equal numbers in each half of the container. Without labels, it is impossible to determine that mixing has occurred. For indistinguishable molecules, the entropy change is 0 $\Delta S = 0$

Both answers seem valid but zero entropy is correct because the molecules cannot be distinguished to restore them to their original locations. However, indistinguishability is even more subtle. Imagine instrumentation (or a computer program) that could observe and follow every molecule in the two containers. The particles are effectively labeled because each location is known at every time. The instrument or computer could then be used to return them to their original positions. However, quantum mechanics postulates that these molecules can never be labeled in this manner. For example, if atom 1 collides with atom 2, the products of the collision produce the expected (1,2) trajectory and also a trajectory (2,1). Indistinguishability is manifested in the need to include both outcomes in the final result.

5.13 Entropy and Information

An increase in states means more system randomness and less order. The probability of finding a state if all are equally probable is

$$p_i = \frac{1}{\Omega}$$

However, systems where the state probabilities are not equal do exist and lead to a more general definition of entropy where the probabilities, not the numbers of states, dominate.

Probabilities can occur in situations outside physics and chemistry. Any of 26 letters could be sent down a communication line. If all were sent with equal probability, the probability of receiving a specific letter would be 1/(26) and the entropy per signal is

$$S = k \ln (26) .$$

Letters do not appear with equal probability. e appears far more often than q so that a signal, given this information on frequency, is less random, i.e., lower entropy. Consider an alphabet with only three letters (a,b,c). Without information, the three letters are sent with probabilities 1/3. However, if the vowel a appears more often, e.g., $p_a = 0.5$ and $p_b = p_c = 0.25$, the entropy drops from $k\ln(3) = 1.10k$ for the random system to

$k\{0.5\ln(0.5) + 0.25\ln(0.25) + 0.25\ln(0.25)\} = 1.04k$. The letter is more likely to be recorded correctly because frequency information is included. The transmission is more ordered.

A DNA molecule is made of four nucleotides (C,G,A,T). If they are present in equal amounts (determined by taking the average from the entire chain), their probabilities are all 1/4 and the entropy for a base is $k\ln(4) = 1.39k$. A more organized system might have different probabilities although the base pair probabilities (A,T) and (C,G) must be equal. Probabilities

$$p_c = p_G = 0.3 \quad p_a = p_T = 0.25$$

give a lower entropy

$$S = k \{2 [0.3 \ln (0.3)] + 2 [0.2 \ln (0.2)] = 1.37 k$$

The decrease in entropy from the most random case in both examples is a measure of external information available to the system. The entropy decrease 0f $1.1k - 1.04k = 0.06k$ for the three letters and $1.39k - 1.37k = 0.02k$ for the more ordered nucleotides defines information. Since this concept is applicable to non-thermodynamic systems, k is dropped and the base 2 logarithm $\log_2$ is used in systems, e.g., computers and communications with binary choices.

External information about the system can be used to lower its entropy. The 50–50 mixture of A and B molecules could be separated into containers with pure A and B if A and B molecules are kept on the left and right sides, respectively. This is done using the following gedanken (thought) experiment. A partition with hole covered by a movable gate divides the container into two parts. A being, often called Maxwell's demon, sits near the gate and opens it only when an A particle can pass through the hole from right to left or a B particle can pass through from left to right. After $2N_a$ gate openings, a mole of A and a mole of B are separated into the left and right containers, respectively.

The separation of the 2 mol requires an entropy decrease of

$$- 2R \ln (2) .$$

The second law, however, states that entropy cannot decrease spontaneously. Entropy of at least $+2R\ln(2)$ must be produced to compensate this decrease. The entropy increase occurs in the Maxwell demon.

As each particle approaches the gate, the demon makes a binary choice – open the gate or do nothing. Since only A and B are present, the probability the gate will be opened is $\frac{1}{2}$. The entropy changes by $k\ln(2)$ for each binary choice. The demon's entropy increases because it no longer stores this bit of information. After 2^{2N} such decisions, the demon's entropy has increased by

$$S = k\ln\left(2^{2N}\right) = 2N_a k\ln(2) = +2R\ln(2)$$

For this reversible system, the decrease in entropy on unmixing is balanced by an equal increase in the entropy of the demon.

A Maxwell demon could also be used to separate hot (high kinetic energy) and cold (low kinetic energy) particles. The gate is opened only when a high energy particle approaches from the right or a low energy particle approaches from the left. A temperature gradient appears and the system entropy decreases. The entropy of the demon increases.

If a temperature gradient is produced without the demon, total entropy decreases in violation of the second law. Consider the case where the two halves of the container are separated by a membrane permeated with protein channel to permit gas flow (Fig. 5.2).

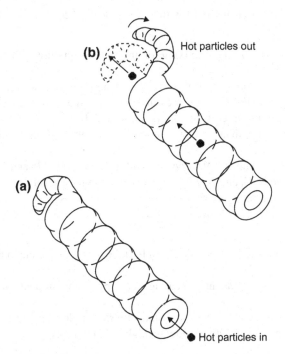

(b)

Hot particles out

(a)

Hot particles in

Fig. 5.2 A protein with prosthetic flap to pass "hot" molecules in one direction

The protein has a prosthetic flap on one end that can be forced open only if a particle in the channel has sufficient kinetic energy. High energy particles can then pass from right to left. A high energy particle on the left cannot open the flap. Even if some kinetic energy was used to open the flap, enough energetic particles would reach the left and raise its temperature.

The temperature gradient created might be tapped to do work (Fig. 5.2). The "cold" side contacts the surroundings to maintain a constant temperature distribution with hot molecules that can lift the flap and reach the "hot" side. This is a perpetual motion machine where heat is converted to work.

This perpetual motion machine must fail because, each time a hot molecule raises the flap, it uses some of that energy to lift the flap. This energy than accumulates in the flap until the flap begins to open and close randomly so that both hot and cold molecules can pass in either direction. No temperature gradient can develop and no work can be done. The increase in entropy of the flap offsets any possible decrease in entropy from separating the hot and cold molecules.

Problems

5.1 Determine the residual entropy for

a. linear CO_2 and b. planar BF_2Cl with bond angles of $120°$.

5.2 Determine the entropy of mixing for an ideal gas mixture with mole fractions $X_A = 0.1$, $X_B = 0.3$, and $X_c = 0.6$.

5.3 Two moles of oxygen at a pressure of 1 atm and 300 K were mixed with 3 mol of nitrogen at 400 K and a pressure of 1 atm and immersed in a heat bath at 300 K to give a mixture at this temperature and 1 atm pressure. Calculate the entropy, energy, and enthalpy for this process. Note: that O_2 and N_2 are diatomic molecules.

5.4 Determine the entropy of a crystal containing $X_{NO} = 0.025$ and $X_{CO} = 0.75$ at absolute zero. The N, C, and O atoms are nearly the same size for crystallization and may crystallize with random orientation.

5.5 A classroom has 100 equivalent seats.

a. What is the "seating" entropy for a single student who sits alone in the same seat all the time?
b. What is the single student's "seating" entropy if the student chooses a seat randomly?
c. What is the total number of seating states for five students who choose from the 100 seats randomly? (all five students might sit on one seat).

5.6 One mole of gaseous A with heat capacity $C_p(A)$ is mixed with 3 mol of gaseous B with heat capacity per mole $C_p(B)$ at constant $P = 1$ atm and $T = 200$ K.

a. Set up the expression for the entropy of mixing per total moles.
b. Determine an expression for the entropy of mixing at 400 K and 1 atm pressure.
c. Give the probability of selecting an A molecule from the mixture.

Chapter 6
Free Energy

6.1 Spontaneity and Entropy

The entropy change for an ideal gas is calculated from temperature, pressure, or volume changes. A phase transition occurs reversibly if the system and surroundings temperatures are equal. Entropy is transferred from surroundings to system on vaporization. Vaporization of a liquid with a boiling point of 300 K and an enthalpy of vaporization of 24,000 J mol^{-1} produces an entropy increase of

$$\Delta S = \frac{H_{vap}}{T} = \frac{24,000}{300} = 80 \, J \, K^{-1} \, mol^{-1}$$

The reversible transfer at the boiling point is an equilibrium. The liquid and vapor remain in equilibrium while the relative fractions of each phase change. Equilibrium is defined by equal and opposite entropies for the system and surroundings.

The free energies incorporate system and surroundings information into a format in system variables. The Gibbs free energy at constant temperature

$$\Delta G = \Delta H - T\Delta S$$

is 0 for the reversible, equilibrium phase transition

$$\Delta G = 24,000 - 300(+ 80) = 0$$

This means (1) the system is at equilibrium for this transition and (2) no useful work can be extracted from this transition.

A truly reversible phase transition is infinitesimally slow since there is no temperature gradient. An irreversible vaporization phase transition occurs when the surroundings temperature is higher than the system temperature and

$$\Delta S_{univ} > 0.$$

M.E. Starzak, *Energy and Entropy*, DOI 10.1007/978-0-387-77823-5_6,
© Springer Science+Business Media, LLC 2010

If $T_{sur} = 301$ K > 300 K, 24,000 J of heat is transferred from the surroundings to the system at 300 K; the reversible path

$$\Delta S_{sys} = \frac{24,000}{300} = +80$$

The entropy change for the surroundings takes place "reversibly" at 301 K

$$\Delta S_{sur} = \frac{-24,000}{301} = -79.7\,J\,K^{-1}$$

The entropy change of the universe for this phase transition is positive

$$\Delta S_{univ} = \Delta S_{sys} + \Delta S_{sur} = 80 + (-79.7) = +0.3\,J\,mol^{-1}\,K^{-1}$$

The universe entropy increases with increased positive temperature difference between surroundings and system and increased irreversibility.

An irreversible condensation also generates entropy in the universe. For a bath at 299 K, the system loses 24,000 J of heat. The system entropy using the reversible path at 300 K decreases by 80 J K^{-1} mol^{-1} (−80). The surroundings entropy adds the 24,000 J on a reversible path at 299 K

$$\Delta S_{sur} = \frac{+24,000}{299} = 80.27\,J\,K^{-1}\,mol^{-1}$$

The system entropy decreases but the entropy of the universe increases for the irreversible condensation

$$\Delta S_{niv} = -80 + 80.27 = +0.27\,J\,K^{-1}\,mol^{-1}$$

Spontaneous (irreversible) phase transitions increase the entropy of the universe. For both vaporization and condensation, heat flows spontaneously from the hotter to the colder region.

The entropy of the universe for the vaporization at 301 K

$$\Delta S_{univ} = \Delta S_{sys} + \Delta S_{sur} = +\Delta S_{sys} + \frac{-\Delta H_{vap}}{T_{sur}}$$

is converted into an energy by multiplying both sides of the equation by the temperature of the surroundings

$$T_{sur}\Delta S_{univ} = T_{sur}\Delta S_{sys} - \Delta H_{vap}$$

The system temperature must equal the surroundings temperature for the phase transition to occur. Replacing T_{sur} by T_{sys} gives the Gibbs free energy in system variables

$$\Delta G = -T_{sur}\Delta S_{univ} = -T_{sys}\Delta S$$

For spontaneous vaporization, $T_{sur} = T_{sys} > T_{bp}$, ΔG is negative $\Delta G = 24,000 - 301(+80) = -80\,J\,mol^{-1}\,K^{-1}$.

For spontaneous condensation, $T_{sur} = T_{sys} < T_{bp}$

$$\Delta G = -24{,}000 - (299)(-80) = -24{,}000 + 23{,}920 = -80 \, J \, mol^{-1}$$

Each spontaneous process has a negative free energy.

An attempt to vaporize the liquid at 299 is impossible as reflected in a positive free energy

$$\Delta G = +24{,}000 - (299)(-80) = -24{,}000 + 23{,}920 = +80 \, J \, mol^{-1} K^{-1}.$$

6.2 The Free Energies and Work

The Gibbs free energy for enthalpy

$$\Delta G = \Delta H - \Delta S$$

Is paralleled by a Helmholtz free energy with internal energy

$$\Delta A = \Delta E - T \Delta S$$

Both energies subtract the reversible heat $T\Delta S$ from the system energies. A reaction with $\Delta H = -10{,}000$ J and $\Delta S = +10$ J K^{-1} mol^{-1} at 300 K gives

$$\Delta G = -10{,}000 - (300)(10) = -13{,}000 \, J \, mol^{-1}$$

By convention, ΔH is negative since the positive energy produced by the reaction is transferred to maintain 25°C. The entropy change is the actual system entropy. $-T\Delta S$ is released to the surroundings if the entropy increases.

The free energy is larger than the enthalpy in this case and measures the work that can be done by this reaction. Both the kinetic energy change for the formation of new bonds in reaction and the potential energy released on the increase in randomization are free to do work on the surroundings.

An exothermic reaction ($\Delta H = -15{,}000$ J mol^{-1}) with a negative entropy change ($\Delta S = -8$ J K^{-1} mol) at 300 K releases

$$\Delta G = -15{,}000 - (300)(-8) = -12{,}600 \, J \, mol^{-1}$$

since some energy released during the bond changes on reaction is now used to order the products and cannot be used to do work.

Entropic heat is transferred to a gas at T_h in a Carnot cycle. The same gas at T_c has an equal and opposite entropy change but less entropic heat. The difference in these two entropic heats is the net work done by the ideal system between the two temperatures:

$$w = q_h - q_c$$

Some of the "latent" heat to increase randomization at the higher temperature is not needed at the lower temperature and can be released as work in the Carnot cycle.

6.3 The Legendre Transform and Thermodynamic Energies

The free energies are state functions made from combinations of other state functions

$$G = H - TS$$
$$A = E = TS$$

Enthalpy is generated as

$$H = E - (-P)V = E + PV$$

In each case, the subtracted product term is an extensive/intensive conjugate pair. In the differential, reversible limit

$$dE = dq_{rev} + dw_{rev} = TdS - PdV$$

$H = E + PV$ is converted to a differential using the product rule

$$dH = dE + PdV + VdP$$

Substituting dE gives

$$dH = TdS - PdV + PdV + VdP = TdS + VdP$$

The independent variables for dH are S and P. If they are constant, $dH = 0$.

The change from the differential expression for E to a differential expression for H is an example of a Legendre transform. The transform changed the internal energy with differential variables S and V to the enthalpy with differential variables S and P. The conjugate variables $-P$ and V in the internal energy expression are multiplied and the product is subtracted from the internal energy to define the enthalpy

$$H = E - (-P)V = E + PV$$

Other energies are defined by subtracting conjugate pairs. The Helmholtz free energy A is generated from the internal energy E in the following steps:

1) a conjugate pair whose product has units of energy (TS) is selected and subtracted from the internal energy to define a new energy, A,

$$A = E - TS$$

2) the new energy, A, is differentiated,

$$dA = dE - TdS - SdT$$

3) dE is substituted into the equation and two identical terms cancel,

$$dA = TdS - PdV - TdS - SdT = -SdT - PdV$$

The Helmholtz free energy is constant if the experimentally controlled thermo-dynamic variables T and V are constant

$$dA = -SdT - PdV = 0$$

The differential Gibbs free energy is defined by subtracting the thermodynamic pair, TS, from the enthalpy

$$G = H - TS$$
$$dG = -SdT + VdP$$

and is constant when T and P are constant.

Internal energy with reversible system work

$$dE = TdS - PdV + \gamma dA$$

gives internal energy in the independent variables S, V, and A. This is Legendre transformed to the new energy

$$E_s = E - \gamma A$$

with differential form

$$dE_s = TdS - PdV - Ad\gamma$$

and independent variables S, V, and γ.

The internal energy with electrical work for transferring charge dq by an electrical potential ψ is

$$dE = TdS - PdV - \psi dq$$

The sign is negative because moving positive charge out of the system with a positive system potential decreases the internal energy of the system. A new free energy, A, is formed by subtracting the product, $-\psi q$,

$$\tilde{A} = E - (-\psi)(q)$$

to give

$$d\tilde{E} = dE - (-\psi)dq - (q)(-d\psi) = TdS - PdV - \psi dq + \psi dq + qd\psi$$
$$= TdS - PdV + qd\psi$$

with independent variables S, V, and P.

The total number of coulombs for a mole of material with charge z is zF, where F is Faraday's constant

$$d\tilde{E} = TdS - PdV + zFd\psi$$

This energy is a convenient choice for laboratory experiments where the intensive electrical potential is held constant. However, the entropy is generally difficult to control and it is expedient to define new free energies with the differential variables (T, V, P) or (T, P, P)

$$d\tilde{A} = -SdT - PdV + zFd\psi$$

$$d\tilde{G} = -SdT + VdP + zFd\psi$$

For dimensional consistency, G, S, and V are all per mole (molar) quantities.

The Legendre transforms follow a consistent pattern. An intensive–extensive product XdY

$$dG = -SdT + VdP + XdY$$

becomes a new energy in the variables T, P, and X by simply reversing the positions of X and Y and negating

$$dG(T,P,X) = -SdT + VdP - YdX$$

6.4 The Mathematical Basis of the Legendre Transform

Each energy can be expanded in its independent variables using partial derivatives. Internal energy expanded in its independent variables S and V

$$dE = \left(\frac{\partial E}{\partial S}\right)_V dS + \left(\frac{\partial E}{\partial V}\right)_T dV$$

is compared with

$$dE = TdS - PdV$$

to define T

$$T = \left(\frac{\partial E}{\partial S}\right)_V$$

and P

$$P = -\left(\frac{\partial E}{\partial V}\right)_S$$

Substituting these partial derivatives for their variables in the Legendre transform makes them slopes in a linear equation

$$H = E + PV = E - \left(\frac{\partial E}{\partial V}\right)_S V$$

$$Y = b + mx$$

The Legendre transform equation describes a linear plot tangent to this curve at V. Subtracting

$$\left(\frac{\partial E}{\partial V}\right)_S (V - 0)$$

from the internal energy removes the V axis and V as an independent variable. The S axis remains and the P which had been a dependent variable for E now becomes an independent variable for the new energy

6.5 The Chemical Potential

The energies are extensive quantities that can be converted into intensive quantities by dividing by moles. For a pure (one component) system, the molar enthalpy, ΔH, is

$$\Delta \overline{H} = \frac{\Delta H}{n}$$

In multicomponent systems, the molar enthalpy is defined as a partial derivative in the moles of one component

$$\overline{H}_i = \left(\frac{\partial H}{\partial n_i}\right)_{S,P,n}$$

for constant independent variables S, p, and n.

dH for one component is

$$dH = \left(\frac{\partial H}{\partial n}\right)_{S,P} dn = \overline{H} dn$$

$$dH(S,P,n) = TdS + VdP + \overline{H} dn$$

The molar Helmholtz free energy

$$\overline{A} = \left(\frac{\partial A}{\partial n}\right)_{T,V}$$

gives

$$dA = -\overline{S}dT - \overline{P}dV + \overline{A} dn$$

The partial molar Gibbs free energy is also called the chemical potential with symbol :,

$$\mu = \overline{G} = \left(\frac{\partial G}{\partial n}\right)_{T,P}$$

For a one component system, the chemical potential is the Gibbs free energy per mole.

The Gibbs free energy in the independent variables, T, P, and n is

$$dG = \left(\frac{\partial G}{\partial T}\right)_{P,n} dT + \left(\frac{\partial G}{\partial P}\right)_{T,n} dP + \left(\frac{\partial G}{\partial n}\right)_{T,P} dn$$

or

$$dG = -SdT + VdP + \mu dn$$

Since any of the energies can be differentiated with respect to n

$$\mu = \left(\frac{\partial G}{\partial n}\right)_{T,P} = \left(\frac{\partial H}{\partial n}\right)_{S,P} = \left(\frac{\partial E}{\partial n}\right)_{S,V} = \left(\frac{\partial A}{\partial n}\right)_{T,V}$$

A Legendre transform for n and μ generates a new free energy, G',

$$G' = G - \mu n$$

or

$$dG' = -SdT + VdP - nd\mu$$

with independent variables T, P, and μ.

6.6 The Gibbs–Helmholtz Equation

The Gibbs free energy change

$$\Delta G = \Delta H - T\Delta S$$

and the Helmholtz free energy change

$$\Delta A = \Delta E - T\Delta S$$

are temperature dependent. A new state function

$$J = G/T$$

is manipulated to produce a linear plot for free energy change with temperature. The function J is differentiated using the product rule,

$$\left(\frac{\partial J}{\partial T}\right)_P = \left(\frac{\partial \frac{G}{T}}{\partial T}\right)_P$$

to give

$$\left(\frac{\partial \frac{G}{T}}{\partial T}\right)_P = \frac{1}{T}\left(\frac{\partial G}{\partial T}\right)_P + G\left(\frac{\partial \frac{1}{T}}{\partial T}\right)_P = -\frac{S}{T} - \frac{G}{T^2}$$

The free energy

$$G = H - TS$$

is substituted to give

$$\left(\frac{\partial J}{\partial T}\right)_P = -\frac{S}{T} - \frac{H - TS}{T^2} = -\frac{H}{T^2}$$

If H is constant, $J = G/T$, is integrated

$$\int_{G_1,T_1}^{G_2,T_2} d\left(\frac{G}{T}\right) = \int_{T_1}^{T_2} \frac{-H}{T^2}dT \frac{G_2}{T_2} - \frac{G_1}{T_1} = \frac{H}{T}\Big|_{T_1}^{T_2} = H\left(\frac{1}{T_2} - \frac{1}{T_1}\right)$$

The same integrals are used for ΔG and ΔH

$$\frac{(\Delta G)_2}{T_2} - \frac{(\Delta G)_1}{T_1} = \Delta H\left(\frac{1}{T_2} - \frac{1}{T_1}\right)$$

Since

$$-\frac{1}{T^2}dT = d\left(\frac{1}{T}\right)$$

the differential form

$$d(G/T) = Hd(1/T)$$

shows that H is now the slope of a plot of G/T versus $1/T$ (Fig. 6.1).
   If H depends on temperature, the integral

$$\int_{T_1}^{T_2} \frac{H(T)}{T^2}dT$$

must be used.

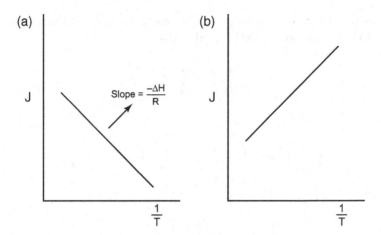

Fig. 6.1 Plots of J versus $1/T$ for a) endothermic and b) exothermic reactions

An exothermic reaction gives a $\Delta G/T$ versus $1/T$ plot with negative slope. As T increases ($1/T$ axis to left), G/T increases. ΔG moves up the energy axis (more positive) indicating less spontaneity in the reaction to products. This is a formal statement of Le Chatelier's principle. An exothermic reaction is driven to reactants at higher temperatures to absorb the extra energy present at the higher temperatures.

6.7 Free Energies and Equilibrium

Free energy measures the drive of a reaction to products. A large negative free energy favors products. However, as products increase, the negative free energy difference decreases. The reaction reaches an equilibrium $\Delta G = 0$ where the pressures or concentrations of reactants and products are constant. The product/reactant pressure ratio when $\Delta G = 0$ is the equilibrium constant K_p. A stable ratio of concentrations defines K_c, the concentration equilibrium constant.

The reaction

$$A \rightleftharpoons B$$

where the free energies of A and B are 20,000 and 30,000, respectively, reaches equilibrium when $A = 0.3$ atm and $B = 0.2$ atm

$$\Delta G = 0.2(30{,}000) - 0.3(20{,}000) = 0$$

and

$$K_p = B^{eq}/A^{eq} = 0.2/0.3 = 2/3$$

Non-equilibrium pressures of A and B will change until the system reaches equilibrium. The molar free energies change with pressure as

$$\overline{dG} = \overline{-SdT} + \overline{V}dP = \overline{V}dP$$

at constant temperature. Molar energies are used. For an ideal gas,

$$V/n = RT/P$$

a change in pressure for 1 mol gives

$$\Delta G = RT \int_{P_1}^{P_2} dP/P$$

Tables of standard free energies list energies for $P = 1$ atm. These free energies are changed to the equilibrium pressure P^{eq}

$$\Delta G = RT \ln (p^{eq}/1)$$

At equilibrium, when the gas pressures are P_A^{eq} and P_B^{eq}, the free energy difference is 0

$$G_B^{eq} - G_A^{eq} = 0$$

$$G_A^o + RTln \left(\frac{P_A^{eq}}{1} \right) = G_B^o + RTln \left(\frac{P_B^{eq}}{1} \right)$$

Collecting terms gives

$$0 = 1G_B^0 - 1G_A^0 + RT \left[\ln \left(P_B^{eq}/1 \right) - \ln \left(P_A^{eq}/1 \right) \right]$$

The ratio of pressure ratios is the equilibrium constant

$$\Delta G^0 = -RT \ln \left(\frac{\frac{P_B^{eq}}{1}}{\frac{P_A^{eq}}{1}} \right) = -RTln(K_P)$$

The equilibrium constant is unitless since each pressure is divided by 1 atm. The ratio gives a free energy correction that brings the standard free energy difference to zero, i.e., equilibrium.

A multicomponent reaction

$$aA + bB = cC + dD$$

with a moles of A, b moles of B, etc., has component free energies $aG_A{}^o$, $bG_A{}^o$, etc. The free energy change to bring each to its equilibrium pressure is

$$-RT\left[c\ln\left(\frac{P_C^{eq}}{1}\right)+d\ln\left(\frac{P_D^{eq}}{1}\right)-a\ln\left(\frac{P_A^{eq}}{1}\right)-b\ln\left(\frac{P_B^{eq}}{1}\right)\right]$$

The free energies with appropriate stoichiometric coefficients and equilibrium pressures are combined as products less reactants. For the standard free energies

$$cG_C^0 + dG_D^0 - aG_A^0 - bG_B^0$$

Including the pressure correction terms and a net free energy of zero at equilibrium gives

$$\Delta G_r^0 = -RT\left(\frac{\left(\frac{P_C^{eq}}{1}\right)^c\left(\frac{P_D^{eq}}{1}\right)^d}{\left(\frac{P_A^{eq}}{1}\right)^a\left(\frac{P_b^{eq}}{1}\right)^b}\right) = -RT\ln\left(K_p\right)$$

The equilibrium pressures of each of the gases are raised to a power equal to its stoichiometric coefficient.

All pressures in the equilibrium constant appear as ratios so the full equilibrium constant is formally dimensionless. If the standard state is 1 atm pressure,

$$K_p = \frac{\left(P_C^{eq}\right)^c\left(P_D^{eq}\right)^d}{\left(P_A^{eq}\right)^a\left(P_B^{eq}\right)^b}$$

The equilibrium constant is constant because the standard free energy difference (all species at 1 atm) is constant.

The free energy to change the pressure from 1 atm to P^{eq} is determined from the ideal gas law. If any reactants or products are either solids or liquids, the change VdP is small and can be ignored. The equilibrium constant includes only the pressures of the gases in the reaction. For the reaction,

$$aA(\text{solid}) + bB(\text{gas}) \rightarrow cC(\text{liquid})$$

the equilibrium constant is

$$K_p = \frac{1}{\left(P_B^{eq}\right)^b}$$

An equilibrium constant for gases A and B that obey the ideal gas law

$$K_p = \frac{P_B}{P_A}$$

is converted to a gas concentration in mol L^{-1} by substituting

$$p_A = \frac{n_A RT}{V} = RT\frac{n_A}{V} = RT[A]$$
$$P_B = RT[B]$$

To give a concentration ratio

$$K_c = \frac{[B]^1}{[A]^1}$$

For the reaction,

$$aA + bB = cC$$

the pressure equilibrium expression

$$K_p = \frac{\left(p_C^e\right)^c}{\left(p_A^e\right)^a \left(p_B^e\right)^b}$$

becomes

$$K_c = \frac{\left(c_C^e\right)^c}{\left(c_A^e\right)^a \left(c_B^e\right)^b}(RT)^{c-a-b}$$

K_c can also be determined directly from a table of free energies at 1 M concentration

$$\Delta G_M^0 = -RT \ln(K_c)$$

6.8 Fractional Concentrations or Pressures

The equilibrium constant relates product and reactant pressures at equilibrium. In situations with total pressure the only observable, the partial pressures are determined from K. The total pressure for the 1:1 reaction

$$A \rightarrow B$$

is p_t and

$$p_B^e = p_t - p_A^e$$
$$K_p^{eq} = \frac{p_t - p_A^e}{p_A^e}$$

and $p_A{}^e$ is

$$p_A^{eq} = \frac{p_t}{1 + K_p^{eq}} K^{eq}$$
$$p_B^e = p_t - p_A^e$$
$$p_B^{eq} = \frac{p_t}{1 + K_p^{eq}}$$

The fractions of each species at equilibrium are the ratios of the equilibrium pressure to the total pressure

$$f_A = p_A/p_t = 1/(1 + K)$$
$$f_B = p_B/p_t = K/1 + K$$

The fractions are generated if A is assigned a "partial pressure" of 1 and B is assigned a "partial pressure" of K. The total pressure in the denominator is the sum of these partial pressures. The numerators are the portions of this total allotted to each species.

The single-site enzyme reaction

$$E + S \rightleftharpoons ES$$

with $K_c = K$ gives ES in terms of E

$$[ES] = K[S][E]$$

Assigning $E = 1$, $[ES] = K[S]$

The concentrations of E and ES relative to E_t, the total enzyme concentration are

$$[E] = \{1/1 + K[S]\} E_t$$
$$[ES] = \{K[S]/1 + K[S]\} E_t$$

An enzyme E with two identical substrate binding sites has four distinct enzyme forms, E, ES, SE, and SES

$$E + S \rightleftharpoons ES, SE + S \rightleftharpoons SES$$

The equilibrium between E and ES with equilibrium constant K_1 is

$$K_1 = \frac{[ES]}{[E][S]} = \frac{[SE]}{[E][S]}$$

for either ES or SE. The concentration of ES or SE is proportional to E

$$[ES] = [SE] = K_1[E][S]$$

The second equilibrium converts either ES or SE into SES,

$$K_1 = \frac{[SES]}{[SE][S]} = \frac{[SES]}{[ES][S]}$$

and is proportional to ES or SE

$$[SES] = K_1[ES][S] = K_1[SE][S]$$

Since ES and SE depend on the concentration of E, SES also depends on E,

$$[SES] = K_1[ES][S] = K_1^2[E][S]^2$$

The total enzyme concentration, E_t, is the sum of these four species,

$$E_t = [E] + [ES] + [SE] + [SES]$$

Substituting

$$E_t = [E] + K_1[E][S] + K_1[E][S] + K_1^2[E][S]^2 = [E](1 + 2K_1[S] + K_1^2[S]^2$$
$$f_E = [E]/E_t = 1/[1 + 2K_1[S] + K_1^2[S]^2]$$

$$[ES] = [SE] = K_1[S][E] = \frac{K_1[S]E_t}{1 + 2K_1[S] + K_1^2[S]^2}$$
$$F_{ES,SE} = K_1[S]/[1 + 2K_1[S] + K_1^2[S]^2]$$

$$[SES] = K_1^2[S]^2[E] = \frac{K_1^2[S]^2E_t}{1 + 2K_1[S] + K_1^2[S]^2}$$
$$F_SES = K_1^2[S]^2/[1 + 2K_1[S] + K_1^2[S]^2]$$

The fractional ratios of each species are determined by selecting $E=1$, $[ES]=[SE] = K_1[S]$, $[ES_2] = K_1^2[S]^2 = K_1^2[S]^2$. The fraction of $[ES]$ using these concentrations is

$$F_{ES} = \frac{K_1[S]}{1 + K_1[S] + K_1[S] + K_1[S]K_1[S]}$$

The fractions of ES_2 and E are, respectively,

$$F_{ES_2} = \frac{K_1^2[S]^2}{1 + K_1[S] + K_1[S] + K_1[S]K_2[S]}$$
$$F_E = \frac{1}{1 + K_1[S] + K_1[S] + K_1[S]K_2[S]}$$

In general, one enzyme state (the free enzyme E in this case) is selected as a reference. The numerator for any other state is determined as the product of all the equilibrium constant–substrate products that lead to that state.

For a complex ESI, the numerator for ESI in the equilibrium sequence

$$E = ES = ESI$$

is $K_s[S]K_I[I]$ where K_s and K_I are the equilibrium constants for binding to S and I sites, respectively.

Problems

6.1 Given that $H = E + PV$ and $G = H - TS$, show $dG = -SdT + VdP$.

6.2 A system initially with $E = 900$, $H = 1500$, $T = 300$ K, $S = 2$, and $P = 1$ atm changes to a new state with $E = 1200$, $H = 2000$, $T = 400$ K, $S = 3$, and $P = 2$ atm. All energy units are joules. Determine the change in the Gibbs free energy for this change of state. Note: the S are absolute entropies.

6.3 The system energy for expanding a surface by dA is proportional to the surface tension, γ, with units of J m^{-2}

$$dG = \overline{-S}dT + \overline{V}dP + \gamma dA - Vdq$$

a. Define a new free energy, G', with independent intensive variables (T, P, γ, and V).

6.4 The normal boiling point of n-hexane is 69°C. Estimate

a. its molar heat of vaporization and
b. its vapor pressure at 60°C.

Chapter 7
Thermodynamic Equations of State

7.1 Maxwell's Relations

An entropy change for the independent variables T and V follows the partial differential path

$$dS = \left(\frac{\partial S}{\partial T}\right)_V dT + \left(\frac{\partial S}{\partial V}\right)_T dV$$

For an ideal gas,

$$\left(\frac{\partial S}{\partial T}\right)_V = \frac{c_V}{T}$$

$$\left(\frac{\partial S}{\partial V}\right)_T = \frac{R}{V}$$

The Helmholtz free energy change

$$dA = \left(\frac{\partial A}{\partial T}\right)_V dT + \left(\frac{\partial A}{\partial V}\right)_T dV$$

is compared to

$$dA = -SdT - PdV$$

so that

$$\left(\frac{\partial A}{\partial T}\right)_V = -S \qquad \left(\frac{\partial A}{\partial V}\right)_T = -P$$

A partial derivative is an operation. A second partial differentiation can use the variable held constant in the first. The final result is independent of differentiation order, For example,

$$\left(\frac{\partial \frac{RT}{V}}{dT}\right)_V = \frac{R}{V} \qquad \left(\frac{\partial \frac{R}{V}}{\partial V}\right)_T = -\frac{R}{V^2}$$

M.E. Starzak, *Energy and Entropy*, DOI 10.1007/978-0-387-77823-5_7,
© Springer Science+Business Media, LLC 2010

or, in reverse order

$$\left(\frac{\partial \frac{RT}{V}}{\partial V}\right)_T = -\frac{RT}{V^2} \qquad \left(\frac{\partial \left(-\frac{RT}{V^2}\right)}{\partial T}\right)_V = -\frac{R}{V^2}$$

In general,

$$\left(\frac{\left(\partial \frac{\partial f}{\partial T}\right)_V}{\partial V}\right)_T = \frac{\partial^2 f}{\partial T \partial V} = \left(\frac{\partial \left(\frac{\partial f}{\partial V}\right)_T}{\partial T}\right)_V$$

For the Helmholtz free energy A

$$\left(\frac{\left(\frac{\partial A}{\partial T}\right)_V}{\partial V}\right)_T = \left(\frac{\left(\frac{\partial A}{\partial V}\right)_T}{\partial T}\right)_V$$

since

$$\left(\frac{\partial A}{\partial T}\right)_V = -S$$

$$\left(\frac{\partial A}{\partial V}\right)_T = -P$$

$$\left(\frac{\partial (-P)}{\partial T}\right)_V = \left(\frac{\partial (-S)}{\partial V}\right)_T$$

$$\left(\frac{\partial (P)}{\partial T}\right)_V = \left(\frac{\partial (S)}{\partial V}\right)_T$$

This Maxwell relation relates the entropy change with volume at constant temperature to a tractable derivative in pressure and temperature.

The equation of state n moles of an ideal gas

$$PV = nRT$$
$$P = nRT/V$$

is substituted in this Maxwell relation

$$\left(\frac{\partial S}{\partial V}\right)_T = \left(\frac{\partial P}{\partial T}\right)_V = \frac{nR}{V}\left(\frac{\partial T}{\partial T}\right) = \frac{nR}{V}$$

to give the result derived previously. dS is

$$dS = \left(\frac{\partial S}{\partial V}\right)_T dV = \frac{nR}{V} dV$$

Maxwell's relation can be used for any equation of state. The entropy change with volume for n moles of gas with equation of state

$$P(V - nb) = nRT$$

is

$$dS = \left(\frac{\partial S}{\partial V}\right)_T dV = \left(\frac{\partial \frac{nRT}{V-nb})}{\partial T}\right) dV = \frac{nR}{V - nb} dV$$

Since

$$\int \frac{dx}{x - b} = \ln(x - b)$$

$$S = \Delta S = \int_{V_1}^{V_2} \frac{nR}{V - nb} dV = R \ln \left(\frac{V_2 - nb}{V_1 - nb}\right)$$

7.2 Gibbs Free Energy

The Gibbs free energy

$$dG = -SdT + VdP$$

in temperature and pressure has the Maxwell relation

$$\left(\frac{\partial \left(\frac{\partial G}{\partial T}\right)_P}{\partial P}\right)_T = \left(\frac{\partial \left(\frac{\partial G}{\partial P}\right)_T}{\partial T}\right)_P$$

with

$$\left(\frac{\partial G}{\partial T}\right)_P = -S$$
$$\left(\frac{\partial G}{\partial P}\right)_T = V$$

the Maxwell relation is

$$\left(\frac{\partial(-S)}{\partial P}\right)_T = \left(\frac{\partial(+V)}{\partial T}\right)_P$$
$$\left(\frac{\partial S}{\partial P}\right)_T = \left(\frac{\partial V}{\partial T}\right)_P$$

The isothermal entropy change with pressure for an ideal gas

$$\left(\frac{\partial S}{\partial P}\right)_T = -\left(\frac{\partial \left(\frac{RT}{P}\right)}{\partial T}\right)_P = \frac{-R}{P}$$

gives

$$dS = \left(\frac{\partial S}{\partial P}\right) dP$$

$$\Delta S = \left(\frac{\partial S}{\partial P}\right)_T dP = -R \int_{P_i}^{P_f} \frac{dP}{P} = -R \ln \frac{P_f}{P_i}$$

The entropy change for the non-ideal gas with equation of state

$$V = RT/P + b$$

from P_i to P_f is

$$\left(\frac{\partial S}{\partial P}\right)_T = -\left(\frac{\partial V}{\partial T}\right)_P$$

$$dS = -\left(\frac{\partial \left[\frac{RT}{P} + b\right]}{\partial T}\right)_P dP = -\frac{R}{P} dP$$

$$\Delta S = -\int_{P_i}^{P_f} \frac{RT}{P} = -RT \ln \left(\frac{P_f}{P_i}\right)$$

The Maxwell relations can combine any two conjugate energy pairs. The independent (differential) variables always appear in the denominator of the derivative.

The Maxwell relation for the enthalpy

$$dH = TdS + VdP$$

is

$$\left(\frac{\partial T}{\partial P}\right)_S = \left(\frac{\partial V}{\partial S}\right)_P$$

This equation is less useful since S must be the independent variable. Enthalpy is transformed to the Gibbs free energy in independent variables T and P to produce a useful Maxwell's relation.

7.3 Other Maxwell's Relations

Maxwell's relations relate entropy to other system variables by coupling $-SdT$ with other conjugate energy pairs. The work required to stretch a rubber band or polymer is proportional to the stretched length of the polymer and the applied external force,

$$dw = +F_{ext}dL$$

which is converted into a reversible work by replacing F_{ext} with the restoring force or tension, τ, of the polymer

$$\tau dL$$

The energy is positive because the reversible work of expansion increases the internal energy of the system. At constant polymer volume

$$dA' = -SdT - PdV + \tau dL = -SdT + \tau dL$$

The unstretched rubber band has disordered molecular polymers that become ordered as the band is stretched to lower the polymer entropy. The Maxwell relation from A'

$$\left(\frac{\partial S}{\partial L}\right)_{V,T} = -\left(\frac{\partial \tau}{\partial T}\right)_{V,L}$$

is used to determine an entropy change using the equation of state

$$\tau = AT(L - L_0)$$

with constant A and L_0

$$\left(\frac{\partial S}{\partial L}\right)_T = -\left(\frac{\partial [AT(L - L_0)]}{\partial T}\right)_L = -A(L - L_0)$$

Integrating from L_0 to L

$$\left(\frac{\partial S}{\partial L}\right)_T dL = A(L - L_0) dL$$

$$\Delta S = -A \int_{L_0}^{L} (L - L_0) dL = -A\frac{(L - L_0}{2}\Big|_{L_0}^{L} = -A\frac{(L - L_0)^2}{2}$$

The entropy decrease with length reflects the ordering of the polymer on stretching.

The reversible path for entropy requires the tension, not the external force. The entropy change defines the reversible heat

$$q_{rev} = T\Delta S = -AT\frac{(L - L_0)^2}{2}$$

The heat released by the band during stretching is easily detected by stretching a rubber band and holding it against your cheek.

An electrical potential (j/C) is multiplied by the charge $dq(C)$ to give the electrical work for the system

$$dG' = -SdT - \psi dq$$

The electrical work is negative because a positive system potential uses system energy to move positive charge out of the system.

The entropy change with charge is determined from the change in potential with temperature

$$\left(\frac{\partial (-S)}{\partial q}\right)_T = \left(\frac{\partial (-\psi)}{\partial T}\right)_q$$

The electrical potential of a battery is measured at a series of temperatures to determine

$$\left(\frac{\partial \psi}{\partial T}\right)_C$$

and the change in entropy for transfer of charge.

7.4 Adiabatic Demagnetization

The heat capacity of a perfect crystal decreases to 0 as the temperature approaches 0 K. The decreased heat capacities of both the system and surroundings at low temperature make further cooling more difficult.

If a magnetic solid is placed in a magnetic field, heat is released from the system as the solid becomes more ordered. If the solid is insulated and the magnetic field is turned off, the system temperature drops during this adiabatic demagnetization as internal energy is converted to random heat of disorder.

Magnetic molecules oriented by a magnetic field H produce a net macroscopic magnetization M of the system.

$$M = \chi_m H$$

χ_m is inversely proportional to the temperature of the system

$$\chi_m = \frac{C'_m}{T}$$

The internal energy change depends on the magnetization change dM

$$dE = TdS - PdV + \mu_0 HdM$$

μ_0, the magnetic permittivity gives the proper units. E is Legendre transformed to free energy in variables T, P, and H,

$$dG = -SdT + VdP - \mu_0 MdH$$

to generate a Maxwell relation

$$\left(\frac{\partial (-S)}{\partial H}\right)_T = \left(\frac{\partial (-\mu_0 M)}{\partial T}\right)_H$$
$$\left(\frac{\partial S}{\partial H}\right)_H = \left(\frac{\partial \mu_0 M}{\partial T}\right)_H$$

Using the equation of state

$$\left(\frac{\partial S}{\partial H}\right)_T = \left(\frac{\partial \left[\frac{\mu_0 C_m H}{T}\right]}{\partial T}\right) = -\frac{C'_m H}{T^2}$$

The entropy change when the field is increased from 0 to H is

$$\Delta S = -\frac{C'_m}{2T^2} \int\limits_0^H H dH = -\frac{\mu_0 C'_m H^2}{2T^2}$$

The entropy decreases with increased magnetic field.

If the system is now insulated and the magnetic field is turned off (adiabatic demagnetization), the internal energy and the temperature decrease

$$dE = CdT = \mu_0 M dH = -\left[C'_m/T\right] H dH$$
$$\int_{T_1}^{T_2} CT dT = \int_H^0 C'_m H dH$$
$$\frac{C}{2}\left[T_2^2 - T_1^2\right] = -\frac{C'_m H^2}{2T}$$

The molecules are free to randomize and the temperature decreases.

7.5 The Lippman Equation

When a potential is applied to mercury in a capillary, the mercury rises or falls in the capillary depending on the polarity. This is electrocapillarity. The phenomenon is caused by a change in the liquid surface tension. Maxwell's relations connect an electrical potential energy

$$- \psi dq$$

to the reversible system work for changing the surface area A with a surface tension, ψ, with units of energy per unit area

$$- \gamma dA$$
$$dG = -\psi dq - \gamma dA$$

The change in surface tension with voltage requires a Legendre transform

$$dG' = +qd\psi - \gamma dA$$

to give the Maxwell relation

$$\left(\frac{\partial \gamma}{\partial \psi}\right)_A = -\left(\frac{\partial q}{\partial A}\right)_\psi = -\sigma$$

σ is the charge per unit area.

The equation of state for γ and ψ is

$$\gamma = -\frac{\overline{C}\psi^2}{2}$$

The derivative is linear

$$\frac{\partial \gamma}{\partial \psi} = -C\psi = -\frac{\partial q}{\partial A} = -\sigma$$

The constant C is the capacitance per unit area

$$C = \sigma/\psi$$

The capacitance is produced by separated charges of opposite polarity on the drop surface and in the solution, respectively, a double layer capacitance. The differential capacitance

$$\overline{C} = \left(\frac{\partial \sigma}{\partial \psi}\right)$$

is identical in this case.

7.6 Thermodynamic Equations of State

The derivative

$$\left(\frac{\partial E}{\partial V}\right)_T$$

is applied to

$$dE = TdS - PdV$$

to give

$$\left(\frac{\partial E}{\partial V}\right)_T = T\left(\frac{\partial S}{\partial V}\right)_T - P\left(\frac{\partial V}{\partial V}\right)_T = T\left(\frac{\partial S}{\partial V}\right)_T - P$$

Since

$$\left(\frac{\partial S}{\partial V}\right)_T = \left(\frac{\partial P}{\partial T}\right)_V$$

the thermodynamic equation of state

$$\left(\frac{\partial E}{\partial V}\right)_T = T\left(\frac{\partial P}{\partial T}\right)_V - P$$

requires only an equation of state with P, T, and V to determine

$$\left(\frac{\partial E}{\partial V}\right)_T$$

For an ideal gas

$$\left(\frac{\partial P}{\partial T}\right)_V = \left(\frac{\partial \frac{nRT}{V}}{\partial T}\right)_V = \frac{nR}{V}$$

$$\left(\frac{\partial E}{\partial V}\right)_T = T\left(\frac{nR}{V}\right) - \frac{nRT}{V} = 0$$

The internal energy of an ideal gas does not change with a change in volume. The van der Waals equation of state for 1 mole of gas

$$\left(P + \frac{a}{V^2}\right)(V - b) = RT$$

$$P = \frac{RT}{V - b} - \frac{a}{V^2}$$

gives

$$\left(\frac{\partial P}{\partial T}\right)_V = \left(\frac{\partial \frac{RT}{V-b}}{\partial T}\right)_V - \left(\frac{\partial \frac{a}{V^2}}{\partial T}\right)_V = \frac{R}{V - b}$$

and

$$\left(\frac{\partial E}{\partial V}\right)_T = \frac{TR}{V - b} - \left[\frac{RT}{V - b} - \frac{a}{V^2}\right] = +\frac{a}{V^2}$$

The energy change depends on the magnitude of a and the inverse square volume and is largest at small volumes when the average distance between molecules is smaller.

The isothermal internal energy change of a van der Waals gas

$$dE = C_v dT + \left(\frac{\partial E}{\partial V}\right)_T dV = \frac{a}{V^2} dV$$

For a volume change from V_i to V_f

$$\Delta E = \int_{V_i}^{V_f} \frac{a}{V^2} dV = -\frac{a}{V} \Big|_{V_i}^{V_f} = -a \left(\frac{1}{V_f} - \frac{1}{V_i} \right)$$

is positive when the volume increases. The potential energy in intermolecular interactions at the smaller volume is released to the system on expansion.

The reversible expansion work uses P_{int},

$$P_{\backslash nt} = \frac{RT}{V - b} - \frac{a}{V^2}$$

$$W = -\int_{V_i}^{V_f} \left[\frac{RT}{V - b} - \frac{a}{V^2} \right] dV = -RT \ln \frac{V_f - b}{V_i - b} + a \left[\frac{1}{V_f} - \frac{1}{V_i} \right]$$

The first term is negative when $V_f > V_i$ since the system is doing work but the negative first term is reduced by the positive second term. This is the energy that is used to separate the molecules as the volume increases.

Substituting the internal energy change and work into the first-law expression

$$q = \Delta E - w = -a \left[\frac{1}{V_f} - \frac{1}{V_i} \right] + RT \ln \frac{V_f - b}{V_i - b} + a \left[\frac{1}{V_f} - \frac{1}{V_i} \right]$$

$$q = RT \ln \frac{V_f - b}{V_i - b}$$

The potential energy released on separating the molecules goes into work, not heat.

7.7 The Joule–Thomson Coefficient

The thermodynamic of state for the change of enthalpy with pressure at constant temperature

$$\left(\frac{\partial H}{\partial P} \right)_T$$

is derived by differentiating the differential enthalpy expression,

$$dH = TdS + VdP$$

$$\left(\frac{\partial H}{\partial P} \right)_T = T \left(\frac{\partial S}{\partial P} \right)_T + V \left(\frac{\partial P}{\partial P} \right) = T \left(\frac{\partial S}{\partial P} \right)_T + V$$

and substituting the Maxwell relation

$$\left(\frac{\partial S}{\partial P}\right)_T = -\left(\frac{\partial V}{\partial T}\right)_P$$

to give a thermodynamic equation of state

$$\left(\frac{\partial H}{\partial P}\right)_T = -T\left(\frac{\partial V}{\partial T}\right)_P + V$$

For an ideal gas,

$$\left(\frac{\partial V}{\partial T}\right)_P = \left(\frac{\partial \frac{RT}{P}}{\partial T}\right)_P = \frac{R}{P}$$

$$\left(\frac{\partial H}{\partial P}\right)_T = \frac{-TR}{P} + V = -V + V = 0$$

The enthalpy of an ideal gas does not change with pressure.
The partial derivative

$$\left(\frac{\partial V}{\partial T}\right)_P$$

is complicated for a van der Waals gas. However, a simpler equation of state with $a = 0$,

$$P(V - b) = RT$$

gives

$$\left(\frac{\partial V}{\partial T}\right)_P = \left(\frac{\partial \left[\frac{RT}{P} + b\right]}{\partial T}\right) = \frac{R}{P}$$

and

$$\left(\frac{\partial H}{\partial P}\right)_T = -\frac{RT}{P} + V = -\frac{RT}{P} + \frac{RT}{P} + b = b$$

For constant b, the enthalpy change of the gas at constant temperature is

$$\Delta H = \int_{P_i}^{P_f} b\, dP = b\,(P_f - P_i)$$

in Latm/mol.

Since the constants a and b make small corrections to the volume and pressure of the gas, a product term with a and b is also small. Expanding factors in the van der Waals equation and eliminating the ab term gives the approximate equation of state

$$PV - bP + \frac{a}{V} = RT$$

$$PV = RT + bP - \frac{a}{V}$$

Since a is small, the V in the a/V term is replaced with its ideal gas value to give an expression for V in terms of P and T

$$RT = P(V - b) - \frac{aP}{RT}$$

The temperature derivative is

$$\left(\frac{\partial V}{\partial T}\right)_P = \frac{\partial\left[\frac{RT}{P} + b - \frac{a}{RT}\right]}{\partial T} = \frac{R}{P} + \frac{a}{RT^2}$$

and

$$\left(\frac{\partial H}{\partial P}\right)_T = V - \frac{RT}{P} - \frac{a}{RT} = \frac{RT}{P} + b - \frac{a}{RT} - \frac{RT}{P} - \frac{a}{RT} = b - \frac{2a}{RT}$$

The Joule–Thomson coefficient for the expansion of a gas at constant enthalpy

$$\mu_{JT} = \frac{\left(\frac{\partial H}{\partial P}\right)_T}{C_P}$$

is

$$\mu_{JT} = \frac{-b + \frac{2a}{RT}}{C_P}$$

At high temperatures, when a/RT is small, $\mu > 0$. The Joule–Thomson inversion temperature ($\mu = 0$) is

$$T = \frac{2a}{Rb}$$

Problems

7.1 The thermodynamic properties of light are determined using the relationship between temperature and light pressure.

a. Use the thermodynamic equation of state

$$\left(\frac{l\partial E}{\partial V}\right)_T = T\left(\frac{\partial P}{\partial T}\right)_V - P$$

If the radiation pressure is related to the internal energy per unit volume

$$(E/V) = \left(\frac{\partial E}{\partial V}\right)_T$$

by the equation of state

$$P = \frac{E}{3V}$$

Show

$$T\left(\frac{\partial P}{\partial T}\right)_V = 4P$$

b. Collect terms and integrate to show that the radiation pressure of light is proportional to the fourth power of the temperature.

7.2 A flexible polymer stretches when a tension τ (energy per unit length) is applied to change the length L. The internal energy

$$dE = TdS + \tau dL$$

a. Determine

$$\left(\frac{\partial E}{\partial L}\right)_S$$

from the internal energy.
b. Use dE to create a free energy with independent variables T and L using the Legendre transformation.
c. Determine the change in entropy with length via Maxwell's relations and the equation of state.

7.3 Given the internal energy

$$dE = TdS + \tau dL$$

develop a thermodynamic equation of state in experimental variables for $(dE/dL)_T$.

7.4 Given the equation of state $PV = RT + BP$ where $B = B(T)$, show

$$\left(\frac{\partial E}{\partial V}\right)_T = \frac{RT^2}{(V-B)^2}\frac{dB}{dT}$$

7.5 Derive an expression for the Joule–Thomson coefficient

$$\mu_{JT} = \frac{\left(\frac{\partial H}{\partial P}\right)_T}{C_p}$$

for a gas with equation of state

$$PV = RT - \frac{aP}{T}$$

7.6 Prove the identity

$$C_P = C_V + T \left(\partial P/\partial T\right)_V \left(\partial V/\partial T\right)_P$$

from $\left(\frac{\partial H}{\partial T}\right)_P = C_V + \left\{\left(\frac{\partial E}{\partial V}\right)_T + P\right\} \left(\frac{\partial V}{\partial T}\right)_P$

by substituting $(\partial E/\partial V)_T$.

7.7 Molecules confined to a surface have an internal energy,

$$dE = TdS + \gamma dA$$

a. Develop a thermodynamic equation of state for $(\partial E/\partial A)_T$.
b. If the equation of state for the system is

$$\gamma = E^{s} \left(1 - T/T_c\right)$$

where E^{s} and T_c are constants, determine the entropy change when the surface is expanded from area A_1 to area A_2, A Legendre transform is needed.

7.8 An electric field, D, will work on a system to create a net orientation of induced dipoles (the polarization P) proportional to the field. The work for polarization is $w = PdD$ and the polarization itself is proportional to the electric field, i.e., $P = \alpha D$, where the constant is temperature independent.

a. Give an internal energy differential expression in the variables S, V, and D.
b. Develop a free energy, Z, in the variables T, P, and D.
c. Determine $(\partial Z/\partial D)_P$.
d. Set up the thermodynamic equation of state for $(\partial E/\partial D)_{P,T}$.

7.9 A one-dimensional lattice gas with N particles free to locate on N sites has a Helmholtz free energy

$$dA = -SdT + \mu dN + \phi dM$$

a. Find

$$\left(\frac{\partial A}{\partial N}\right)_{T,M} =$$

$$\left(\frac{\partial \mu}{\partial M}\right)_{N,T} =$$

Chapter 8
Chemical Potentials in Solution

8.1 Chemical Potentials for Ideal Solutions

The free energy change per mole (the chemical potential) for a mole of ideal one component gas at a constant temperature is

$$d\mu = \frac{dG}{n} = \frac{\frac{nRT}{P}dP}{n} = \frac{RT}{P}dP = RTd[\ln(P)]$$

If the gas is in equilibrium with pure liquid, the pressure is the vapor pressure, p^o, of that liquid. For equilibrium, the chemical potentials of gas and liquid must be equal

$$\mu g = \mu_l = RT \ln \left(p^o \right)$$

so that vapor properties give thermodynamic information on liquids and solutions.

If the vapor pressure and chemical potential of the vapor change, the chemical potential of the liquid changes to maintain equilibrium

$$d\mu_{vap} = d\mu_{liq}$$

The vapor pressure for a liquid component changes when it is mixed with other liquids. An ideal mixture obeys Raoult's law. For a mixture of A and B with mole fractions X_A and X_B, respectively,

$$p_A = X_A p_A^o \quad p_B = X_B p_B^o$$

Since the pure vapor pressures are constant, the chemical potential changes only if the mole fraction changes

$$d\mu_A(\text{liq}) = d\mu_A(\text{vap}) = RTd[\ln \left(X_A p_A^o \right)]$$
$$= RTd\ln \left(X_A \right) + RTd[\ln \left(p_A^o \right)] = RTd\ln \left(X_A \right)$$

For the ideal solution, the chemical potential change for B is

$$d\mu_B = RTd\ln \left(X_B \right)$$

M.E. Starzak, *Energy and Entropy*, DOI 10.1007/978-0-387-77823-5_8,
© Springer Science+Business Media, LLC 2010

The chemical potential of water in aqueous solutions depends on its mole fraction,

$$d\,\mu_{\text{liq}}\,(\text{H}_2\text{O}) = RTd\left[\ln\left(X_{\text{H}_2\text{O}}\right)\right]$$

because it relates to vapor pressure through Raoult's law. However, since pressure is directly proportional to concentration through the ideal gas law

$$P = (n/V)RT = cRT$$

the differential chemical potential is also

$$d\,\mu = RTd\ln\{c\}$$

This equation is generally used with solutes that have negligible vapor pressure, e.g., ions.

Each ion in a salt solution has its own chemical potential

$$d\,\mu[\text{Na}] = RTd\ln[\text{Na}^+]$$
$$d\,\mu(\text{Cl}) = RTd\ln[\text{Cl}^-]$$

These potentials sum to give the total chemical potential for the salt as solute

$$d\,\mu = d\,\mu(\text{Na}) + d\,\mu(\text{Cl}) = RTd\,\ln[\text{Na}] + RTd\,\ln\text{Cl} = RTd\ln\,([\text{Na}][\text{Cl}])$$

For $CaCl_2$, the chemical potential for one mole of the salt is the sum of the chemical potentials for one mole of Ca^{2+} and the free energy for two moles of Cl^-

$$d\,\mu = 1RTd\ln[\text{Ca}^{+2}] + 2RTd\ln[\text{Cl}^-] = RTd\ln\left(\left[\text{Ca}^{+2}\right]\left[\text{Cl}^-\right]^2\right)$$

Since the anions and cations always appear together, a mean ion concentration is defined as

$$d\,\mu = RTd\ln\left(\left[\text{Ca}^{2+}\right]\left[\text{Cl}^-\right]^2\right) = RTd\ln\left(c_{\pm}^3\right)$$

8.2 Fugacity

The chemical potential

$$d\,\mu = RTd\ln[P]$$

is based on the ideal gas equation. Real gases differ from ideal gases and the chemical potential for such real gases might be developed in several ways.

(1) An improved equation of state is used in conjunction with

$$d\mu = \overline{V}dP$$

to produce a new function for the chemical potential.

(2) The chemical potential is determined experimentally for different pressures and used to define a corrected pressure or fugacity, f, that gives the proper chemical potential in the logarithmic equation, i.e.,

$$d\mu = RTd[\ln(f)]$$

(3) An improved equation of state is used to produce a logarithmic equation in fugacity instead of pressure. A change in chemical potential is

$$\Delta\mu = RT\ln\left(\frac{f_f}{f_i}\right)$$

The non-ideal equation of state

$$P(\overline{V} - b) = RT$$

for one mole of gas defines a system volume

$$\overline{V} = RT/P + b$$

As pressure approaches 0, the first term dominates and the gas behaves like an ideal gas

$$f_i = P_i \text{ as } P_i \rightarrow 0$$

The chemical potential for the non-ideal gas is also expressed in terms of fugacity for higher pressures

$$d\mu = RTd[\ln(f)]$$

Chemical potential as VdP gives

$$d\mu = \frac{RT}{P}dP + bdP$$
$$\Delta\mu = RT\ln(P/P_i) + b(P - P_i)$$

The two equations are equated

$$\Delta\mu = RT\ln\frac{f}{P_i} = RT\ln\frac{P}{P_i} + b(P - P_i)$$

The limiting pressure, P_i, cancels since $f = P$ at low pressures and

$$RT \ln f - RT \ln P_i = RT \ln P - RT \ln P_i + b (P - P_i)$$

to give

$$\ln f = \ln P + \frac{bP}{RT}$$

or

$$f = Pe^{\frac{bP}{RT}}$$

Because b is small, the fugacity is close to the pressure for this equation of state. The exponential is expanded in a Taylor series

$$e^x = 1 + x + \frac{x^2}{2!} + \dots$$

to give

$$f = P \left(1 + \frac{bP}{RT} + \dots \right) = P + \frac{bP^2}{RT} + \dots$$

8.3 Activity

A solute in dilute solution has a chemical potential
$$d\mu = RTd \ln (c)$$
where c is the molar concentration. The corrected chemical potential for real solutions uses an activity a in place of the molar concentration

$$d\mu = RTd \ln (a)$$

Activity and molarity are related through an activity coefficient

$$a = \gamma c$$

For example, a 0.02 M solution with an activity coefficient of 1.1 at this concentration has an activity

$$a = \gamma c = 1.1(0.02) = 0.022$$

for a more accurate chemical potential.

A salt dissociates into its constituent ions in solution. Each ion then has a concentration and activity that are related. The mean concentration is a geometric mean since the logarithmic chemical potentials add. The mean concentration for salts, $C_{z-}^{z+}A_{z+}^{z-}$, e.g., $Ca_1^{2+}Cl_2^-$, with z_- moles of C and z_+ moles of A

$$c_{\pm}^z = c_C^{z-} c_A^{z+}$$

The mean activity and mean activity coefficient are

$$a_{\pm}^z = a_C^{z-} a_A^{z+}$$

and

$$\gamma_{\pm}^z = \gamma_C^{z-} \gamma_A^{z+}$$

respectively.

For $CaCl_2$,

$$\gamma_{\pm}(CaCl_2) = \left[\gamma_+^1(Ca)\gamma_-^2(Cl)\right]^{1/3}$$

8.4 Partial Molar Quantities

The chemical potentials for a two component system are more complicated for non-ideal solutions. However, a differential path gives the total change in free energy

$$dG = \left(\frac{\partial G}{\partial n_A}\right)_{n_B} dn_A + \left(\frac{\partial G}{\partial n_B}\right)_{n_A} dn_B$$

$$\mu_A = \left(\frac{\partial G}{\partial n_A}\right)_{n_B} \qquad \mu_B = \left(\frac{\partial G}{\partial n_B}\right)_{n_A}$$

The free energies can now be different for different pairs (n_A, n_B). The total differential free energy is

$$dG = \mu_A dn_A + \mu_B dn_B$$

Total volumes depend on partial molar volumes, the volumes per mole of each species for a given composition. If the volumes of one mole of A and B in an ideal solution were 29 and 30 cm^3/mol, respectively, a mixture with one mole of each has total volume

$$V_t = 1\,mol\,A\left(29\frac{cm^3}{mol}\right) + 1\,mol\,B\left(\frac{30\,cm^3}{mol}\right) = 59\,cm^3$$

This is valid when the volumes per mole are constant for all combinations. More generally,

$$dV_t = \left(\frac{\partial V}{\partial n_A}\right)_{n_B} dn_A + \left(\frac{\partial V}{\partial n_B}\right)_{n_A} dn_B$$

If the partial molar volumes are known for a given n_A and n_B, the change in total volume on addition of incremental amounts of A and B is known. Larger additions change the composition and the partial molar volumes so detailed data or an equation describing the system volumes are needed for a total volume measurement.

8.5 Euler's theorem

Euler's theorem serves as a bridge from differential to macroscopic changes in partial molar systems. The volume change for a binary system

$$dV_t = \left(\frac{\partial V_t}{\partial n_A}\right)_{n_B} dn_A + \left(\frac{\partial V_t}{\partial n_B}\right)_{n_A} dn_B$$

has two extensive variables, n_A and n_B, as the independent (differential) variables. Differential changes with the same proportions as the macroscopic n_A and n_B produce no change in the overall composition of the solution. For example, adding $dn_A = 0.02$ and $dn_B = 0.03$ to solution with $n_A = 2$ and $n_B = 3$ and mole fractions $X_A = 0.4$ and $X_B = 0.6$ produces no change in the composition of the solution expressed as mole fraction,

$$X_A = \frac{0.2 + 0.02}{0.2 + 0.02 + 0.3 + 0.03} = 0.4$$

Additions of 0.04 and 0.06 for A and B, respectively, also keep $X_A = 0.4$. The macroscopic ratio of $n_A/n_B = 2/3$ is maintained for the differential changes, e.g., $0.04/0.06 = 2/3$. In fact, the "differential" changes could be increased to 2 and 3 (ratio 2/3) from 0.2 and 0.3 without changing the composition

$$X_A = 2.2/(2.2 + 3.3) = 0.4$$

Euler's theorem states that the differential equation *with extensive independent variables* is integrated by removing the differential d

$$V_t = \bar{V}_A n_A + \bar{V}_B n_B$$

The differential free energy

$$dG = \mu_A dn_A + \mu_B dn_B$$

becomes

$$G = \mu_A n_A + \mu_B n_B$$

The chemical potentials must be known for the particular composition defined by n_A and n_B.

Since S and V are extensive variables, Euler's theorem states that

$$dE = TdS - PdV$$

becomes

$$E = TS - PV$$

8.6 Determining Partial Molar Quantities

The total volume of a binary mixture of n_A moles of A and n_B moles of B is measurable This total volume is plotted against mole fraction (composition). Tangents at specific mole fractions give the partial molar volumes for that composition.

The integrated partial molar volume equation

$$V = \bar{V}_A n_A + \bar{V}_B n_B = \left(\frac{\partial V}{\partial n_A}\right)_{n_B} n_A + \left(\frac{\partial V}{\partial n_B}\right)_{n_A} dn_B$$

is converted to mole fractions

$$<V> = \frac{V_t}{n_A + n_B} = \bar{V}_A X_A + \bar{V}_B X_B$$

Since $X_B = 1 - X_A$,

$$V = \bar{V}_A X_A + \bar{V}_B (1 - X_A) = (\overline{V}_A - \bar{V}_B) X_A + \bar{V}_B$$

This linear equation gives $<V> = \bar{V}_B$ when $X_A = 0$ and $<V> = \bar{V}_A$ when $X_A = 1$. A tangent to the plot at any mole fraction gives the partial molar volumes of B and A as intercepts of $X_A = 1$ and $X_A = 0$ respectively (Fig. 8.1).

8.7 The Gibbs–Duhem Equation

Euler's theorem converts a differential equation in extensive independent variables

$$dV_t = \bar{V}_A dn_A + \bar{V}_B dn_B$$

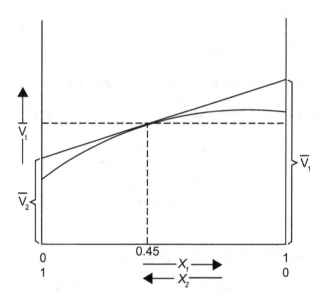

Fig. 8.1 $<V>$ versus X for partial molar volumes

into a macroscopic equation

$$V_t = \bar{V}_A n_A + \bar{V}_B n_B$$

This equation can be differentiated to give four terms

$$dV_t = \bar{V}_A dn_A + n_A d\bar{V}_A + \bar{V}_B dn_B + n_B d\bar{V}_B$$

The formal differential path for total volume is also correct and includes only dn_A and dn_B, differentials

$$dV_t = \bar{V}_A dn_A + \bar{V}_B dn_B$$

The two "correct" equations are reconciled only if

$$0 = n_A d\bar{V}_A + n_B d\bar{V}_B$$

This is an example of a Gibbs–Duhem equation. The partial molar volumes in solution are not independent. If $d\bar{V}_A$ is changed, $d\bar{V}_B$ must change to maintain the equality

$$d\bar{V}_B = -\left(\frac{n_A}{n_B}\right) d\bar{V}_A$$

The internal energy with extensive independent variables

$$dE = TdS - PdV + \mu_A dn_A + \mu_B dn_B$$

becomes

$$E = TS - PV + \mu_A n_A + \mu_B n_B$$

Differentiation gives

$$dE = TdS + SdT - PdV - VdP + \mu_A dn_A + n_A d\mu_A + \mu_B dn_B + n_B d\mu_A$$

and the Gibbs–Duhem equation

$$SdT - VdP + n_A d\mu_A + n_B d\mu_B = 0$$

At constant temperature and pressure, the equation becomes

$$n_A d\mu_A + n_B d\mu_B = 0$$

or

$$n_A d\mu_A = -n_B d\mu_B$$
$$d\mu_B = -(n_A/n_B)\, d\mu_A$$

A chemical potential change for A produces a change in the differential chemical potential for B.

Chapter 9
Phase Equilibria and Colligative Properties

9.1 Chemical Potential Balance Equations

The chemical potential of a system is zero at equilibrium. To maintain equilibrium, a change in T, for example, must be balanced by a change in P, to keep the net chemical potential change equal to zero. For two phases, the chemical potential of any component must be equal. A differential change $d\mu_1$ in phase 1 must be balanced by an equal change, $d\mu_2$, in phase 2. These one and two phase balance equations are used to develop a broad range of equations, the equations for colligative properties.

The Gibbs free energy for a single phase is the sum of intensive–extensive pairs of thermodynamic variables,

$$dG = -SdT + VdP + \sum_{i=1}^{N} \mu_i dn_i - \psi dq + \gamma dA + \dots$$

For a one component system, the free energy G is the chemical potential

$$d\mu \equiv \frac{dG}{n}$$

For chemical potential, the partial molar entropy and volume are used

$$d\mu = \overline{S}dT + \overline{V}dP + \dots$$

The electrical energy

$$+ qd\psi$$

is convenient since a mole of charge has zF coulombs.

$$+ qd\psi = zFd\psi$$

For one phase systems, energy changes are balanced to keep the free energy change zero. For two-phase system, equilibrium between the is maintained by balancing chemical potentials in both phases

$$d\mu(\text{phase 1}) = d\mu(\text{phase 2})$$

M.E. Starzak, *Energy and Entropy*, DOI 10.1007/978-0-387-77823-5_9,
© Springer Science+Business Media, LLC 2010

The chemical potential, not the extensive free energy, is used for the two-phase equilibria.

9.2 The Barometric Equation

The pressure of atmospheric gases decreases exponentially with increasing height h above ground level when temperature is constant. Equilibrium is maintained at all heights to prevent a net motion of gas from one layer to the next under a free energy gradient. Two opposing energies create this equilibrium. The mass m of each particle is directed downward by an energy difference $mgdx$. The same mass is directed upward by a pressure difference.

The free energy change for a mole of particles with pressure P at x

$$\overline{V}dP$$

$$\overline{V}dP = \frac{RT}{P(x)}dP(x)$$

($x = 0$ is ground level) is converted to a free energy per particle ($k = R/N_a$)

$$kTdP(x)/P(x)$$

This change is balanced by the potential energy change

$$d\,[PE(x)] = mgdx$$

where g is the acceleration of gravity.

If P and x are the only independent variables, the change in chemical potential is the sum of these two potential energies. At equilibrium

$$0 = kTdP(x)/P(x) + mgdx$$

At $x = 0$, $P = P_0$, the pressure at ground level. $P(h)$ is the pressure at height h

$$kT\int_{P_0}^{P}\frac{dP}{P} = \int_{0}^{h} mgdx$$

$$P(h) = P_0 e^{-\frac{mgh}{kT}}$$

The exponential ratio of energy mgh to kT is a Boltzmann factor for the probability of finding a particle at height h. kT is a thermal energy that serves as a measure of the energy available to the particle at that temperature.

The barometric formula is also written using the molecular weight M (the mass of a mole of particles) and R (energy mol^{-1} K^{-1})

$$P = P_0 e^{-\frac{Mgh}{RT}}$$

9.3 Sedimentation

Since P is directly proportional to the moles (n) or number of particles (N),

$$\frac{P(h)}{P_0} = \frac{N(h)}{N_0} = e^{\frac{-mgh}{kT}}$$

the barometric equation also predicts the number of particles at each height.

The equation in m and k is valid for single particles. The particle might be a pollen grain or latex sphere. The equation provides an excellent method for determining k. For particles in water, the buoyant mass

$$m' = m(1 - \rho Vg)$$

with the density of water ρ is the density of water and the volume per gram of the particle V_g gives the "air" mass of the particle less the mass of an equal volume of water. A particle with negative buoyant mass floats, while one with positive buoyant mass sinks but also rises into solution via the thermal motions in the water to heights h above the bottom of the tank ($h = 0$).

An experimentalist measures the number of particles at each height in a given cross-sectional area to prepare a logarithmic plot

$$\ln[n(h)] = \ln[n_0] - \frac{m'g}{kT}h$$

that is linear with slope $m'g/kT$ to determine k. Since

$$N_a = R/k$$

Avogadro's number is determined accurately.

The experiments to determine N_a in this manner had far reaching implications because they established that thermodynamic properties were indeed the statistical behavior of a large number of particles. Although the experiments were performed on visible particles, microscopic particles like atoms and molecules followed the same rules.

The effective acceleration of gravity is increased significantly by spinning it in a centrifuge. Particles of different mass then collect at different "heights," i.e., radii, in the centrifuge. For a radial frequency $\omega = 2\pi$ (rotations per second), the acceleration of gravity produced by the centrifuge at a radius r is

$$g' = \omega^2 r$$

g' replaces g and the centrifuge gives a distribution in radius r

$$N(r) = N(r_0)\exp(-m'g'r/kT)$$

9.4 Gibbs-Helmholtz Equation and Equilibrium

The Gibbs–Helmholtz equation (Chapter 7)

$$d\left(\frac{\Delta G}{T}\right) = +\Delta H d\left(\frac{1}{T}\right)$$

determines the change in free energy with temperature.

The free energy change for reactants and products in their standard states defines an equilibrium constant

$$\Delta G^0 = -RT \ln(K_p)$$

A temperature-induced free energy means an equilibrium constant change. Substituting for the free energy

$$d\left(\frac{\ln(K_p)}{RT}\right) = -\frac{\Delta H}{R} d\frac{1}{T}$$

A plot of $\ln(K_p)$ versus $1/T$ is linear with a slope

$$\text{slope} = -\frac{\Delta H}{R}$$

An endothermic reaction ($\Delta H < 0$) produces a negative slope (Fig. 9.1a). The equilibrium constant increases with increasing temperature. This is a quantitative statement of Le Chatelier's principle which states an endothermic reaction goes toward products at higher temperature since the reaction absorbs some of the energy to mitigate the applied force (the temperature rise). An exothermic reaction with a positive slope lowers the equilibrium constant as temperature is increased.

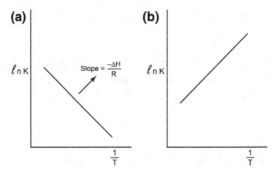

Fig. 9.1 Ln($K/ln(K)$) versus $1/T$ plots for (**a**) an endothermic reaction and (**b**) an exothermic reaction

9.5 Osmotic Pressure

A component in two different phases is equilibrated when its chemical potentials in those phases are equal. A change of chemical potential in one phase is balanced by an equal change in the second phase

$$d\mu_1 = d\mu_2$$

A semipermeable membrane that permits only water flow is a one component, two-phase system. Solute cannot cross the membrane and cannot equilibrate its chemical potentials. Equilibrium for two aqueous solutions (phases) separated by the semipermeable membrane is the equilibrium for water alone.

If pure water is separated from a water solution by the membrane, the mole fraction of pure water ($X = 1$) is higher than the mole fraction of the water in the solution

$$RTd\ln(X_1) + \overline{V}_1 dP_1 = RTd\ln(X_2) + \overline{V}_2 dP$$
$$\mu(X = 1) = RT\ln(1) = 0$$

while

$$\mu(X) = RT\ln(X) < 0$$

In the absence of a counter energy, water flows to the solution. For equilibrium, this driving energy is balanced by a hydrostatic pressure head, the osmotic pressure,

$$\pi = p_2 - p_1$$

The general chemical potential equation

$$d\mu = -\overline{S}dT + \overline{V}dP + RTd\ln X + zF\psi + \dots$$

reduces to two potentials for each phase

$$d\mu_1 = d\mu_2$$
$$\overline{V}_1 dP_1 + RTd\ln(X_1) = \overline{V}_2 dP_2 + RT\ln(X_2)$$

The mole fractions are usually selected for water "concentration"; a molar concentration for solvent is less convenient (Fig. 9.2).

When both baths are pure water, equilibrium requires equal hydrostatic pressures ($P_1 = p_2 = p$). This equilibrium is reversibly changed with increments of solute in Bath 2 that cause incremental pressure changes. An integration sums the reversible steps. The limits of integration are

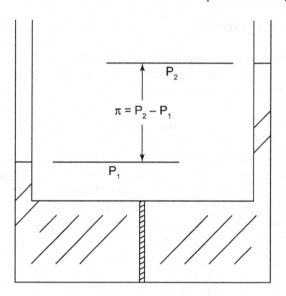

Fig. 9.2 An osmotic pressure defined by the height difference of liquid

Bath 1 (pure water) P to $P_1, X = 1$ to $X = 1$
Bath 2 (solution) P to $P_2, X = 1$ to X

$$\int_P^{P_1} V_1 dP_1 + \int_1^1 RT \ln (X_1) = \int_P^{P_2} V_2 dP_2 + \int_1^X RT \ln (X_2)$$

The concentration integral for phase 1 (pure water) is 0. The remaining three integrals give

$$\bar{V}_1 \int_P^{P_1} dP_1 = \bar{V}_2 \int_P^{P_2} dP_2 + RT \int_{X=1}^X d \ln X \bar{V}_1 (P_1 - P) = \bar{V}_2 (P_2 - P) + RT \frac{\ln (X)}{1}$$

For dilute solutions, the partial molar volume for the pure water and the solution are almost equal because the small amount of solute has little effect on the structure of the water so $\bar{V}_1 = \bar{V}_2 = \bar{V}$

$$\bar{V} (P_1 - P_2) = RT \ln X$$

The pressure difference is the osmotic pressure

$$\pi = P_2 - P_1$$

Dividing each side of the equation by $\overline{V}$ gives

$$\pi = -\frac{RT}{\overline{V}} \ln X$$

The sum of the mole fractions $X_W + X_S = 1$ so

$$X_w = 1 - X_s$$

and

$$\ln X_w = \ln(1 - X_s) \approx -X_s$$

Since

$$X_s = \frac{n_s}{n_s + n_w} \approx \frac{n_s}{n_w}$$

the osmotic pressure is

$$\pi = RT\frac{n_w}{\overline{V}}\frac{n_s}{n_w} = RT\frac{n_s}{V} = cRT$$

When the baths have molar concentrations c_1 and c_2, respectively, the osmotic pressure is

$$\pi = (c_2 - c_1)\,RT$$

The integrations to obtain the expression for the osmotic pressure started from an equilibrium with water in both baths. Solute was added to the right chamber and water flowed to the right to dilute the solution. The osmotic pressure equation describes only the final situation where the pressure has developed and the solution has a different concentration to maintain the equilibrium.

A more practical way to measure osmotic pressure uses external pressure to balance the water flow to solute so that no water flows and the concentration is constant. When $P_{ext} = \pi$, no water can flow to dilute the solution in bath 2.

If the external pressure exceeds the osmotic pressure, water is forced from the solution to the pure water. This reverse osmosis is used to prepare pure water.

9.6 Molecular Weight Measurements

The osmotic pressure of a dilute solution is directly proportional to the molar concentration and large enough for convenient measurement. For example, a 0.01 M solution opposing pure water has an osmotic pressure

$$\pi = cRT = \left(0.01\,\mathrm{mol\,L^{-1}}\right)0.082\frac{\mathrm{Latm}}{\mathrm{mol\,K}}\right)(300\,\mathrm{K}) = 0.246\,\mathrm{atm}$$

One atmosphere supports a 33 foot column of water.

The osmotic pressure is proportional to c, the molar concentration. For g grams of a solute with a molecular weight M in a volume V of solution, the molar concentration is

$$c = \frac{g/M}{V} = G/M$$

$$M = G/c$$

For example, if 0.001 g of a protein is dissolved in 1 mL of water,

$$G = \frac{0.001\,\text{g}}{1\,\text{mL}}\,\frac{10^3\,\text{mL}}{\text{L}} = 1\,\text{gL}^{-1}$$

If the observed osmotic pressure is 0.0025 atm at 300 K, the concentration is (RT $= 25$ Latm)

$$0025 = cRT = 25c$$

$$c = 0.0001$$

the molecular weight of the protein is

$$M = G/c = 1/0.0001 = 10{,}000\,\text{g mol}^{-1}$$

The osmotic pressure counts total solute. A weak acid, HA, with initial concentration c_t dissociates to give the following concentrations

$$\text{HA} = c_t(1 - \alpha)$$

$$\text{H}^+ = \text{A}^- = \alpha c_t$$

$$c_t = [\text{HA}] + [\text{H}^+] + [\text{A}^-] = c(1 - \alpha) + \alpha c + \alpha c = c(1 + \alpha)$$

For a weak electrolyte that dissociates partially into n ions, α is determined from

$$1 + (n - 1)\alpha = \frac{\pi}{c_t RT}$$

A small ion that dissociates from a protein on dissolution has a major impact on molecular weight measurements for that protein. If the protein of 10,000 g/mol had one ionizable Na^+ ion, the osmotic pressure would double since the concentration of solute ions doubles. This doubled concentration gives half the protein molecular weight,

$$M = G/c = 1/0.002 = 5000\,\text{g mol}^{-1}$$

The observed molecular weight is actually an average where, to a first approximation, the weights of the metal ions are 0. For a protein salt P^{3-} with $M = 10{,}000$ and three Na^{3+} (four ions total), the average observed molecular weight is

$$< M > \approx 10{,}000/4 = 2500$$

9.7 The Electrochemical Potential

Solutions are electroneutral. An electrical potential is generated only by a spatial separation of cations and anions. This separation becomes possible when two electroneutral salt solutions of concentrations c_1 and c_2 are separated by a membrane permeable only to cations or anions. A K^+ permeable membrane permits K^+ ions to move down their concentration gradient through the membrane separating it from its counterion (Cl^-). The system equilibrates when the electrical potential free energy balances the concentration free energy for the permeable ion.

The electrochemical potential is

$$d\mu_i = RTd\ln(c_K) + zF\psi$$

K^+ equilibrates between the two phases

$$d\mu_1 = d\mu_2$$
$$RT\,d\,\ln[K]_1 + zFd\psi_1 = RT\,d\ln[K]_2 + zFd\psi_2$$

The initial K^+ concentrations and potentials are

$$[K]_1 = [K]_2 = [K]_0 \qquad \psi_1 = \psi_2 = 0$$

As the concentrations in the baths are changed reversibly to their final values of $[K]_1$ and $[K]_2$, the potentials for each bath change to their final non-zero values of ψ_1 and ψ_2, i.e.,

$$\int_{[K]_0}^{[K]_1} RT\,d\ln c + zF\int_0^{\psi_1} d\psi = \int_{[K]_0}^{[K]_2} RT\,d\ln c + zF\int_0^{\psi_2} d\psi$$

$$RT\,(\ln[K]_1 - \ln[K]_0) + zF\psi_1 = RT(\ln[K]_2 - \ln[K]_0) + zF\psi_2$$

$$RT\ln[K]_1 + zF\psi_1 = RT\ln[K]_2 + zF\psi_2$$

The electrical potential difference is

$$\Delta\psi = \psi_2 - \psi_1 = \frac{RT}{zF}\ln[K]_1 - \frac{RT}{zF}\ln[K]_2 = -\frac{RT}{zF}\ln\frac{[K]_2}{[K]_1}$$

For K^+, $z = +1$. The minus signifies the opposition of the two gradients. A larger concentration of cation in bath 2 produces a larger positive potential in bath 1.

The ratio RT/F is

$$\frac{RT}{F} = \frac{8.31\,\mathrm{J\,mol^{-1}K^{-1}}\,298\,\mathrm{K}}{96{,}500} = 0.0257\,\mathrm{V} \approx 25\,\mathrm{mV}$$

The potential difference for a $[K]_1 = 1$ M and $[K]_2 = 2$ M is

$$\Delta\psi = -\frac{RT}{F}\ln\frac{[K]_2}{[K]_1} = -0.0257\ln\frac{2}{1} = -0.018\,\mathrm{V}$$

A voltmeter measures only the electrical potential difference between the solutions, not absolute potentials. If the left solution is assigned a reference potential of 0 volts, the right solution potential is negative

$$\Delta\psi = \psi_2 - \psi_1 = (-17.6\,\text{mV}) - 0 = -17.6\,\text{mV}$$

Membranes permeable to ions might be used to tap energy where river water mixes with sea water. Fresh water from a river with a low ionic concentration flows along one face of a membrane array. Seawater flows by the opposite face to produce an electrical potential difference across the membrane.

A glass electrode pH meter behaves as if it is permeable only to protons. An acid concentration gradient across the glass bulb produces an electrical potential difference that can be measured. If the internal acid concentration is known, the external sample solution concentration is determined. The pH meter measures small concentrations because the potential is determined by concentration ratios, not absolute concentrations. A concentration ratio of 0.01/0.02 gives the same electrical potential as a ratio $1\times10^{-8}/2\times10^{-8}$.

9.8 The Clapeyron Equation

A liquid–solid equilibrium for one component equilibrates when

$$d\mu_s = d\mu_l$$

A change in pressure changes the equilibrium melting temperature where the phases coexist. Both phases have the same temperature and pressure. The chemical potential balance equation

$$-\overline{S}_s dT_s + \overline{V}_s dP_s = -\overline{S}_{\text{liq}} dT_{\text{lit}} + \overline{V}_{\text{liq}} dP_{\text{liq}}$$

is rearranged

$$\left(-\overline{S}_s + \overline{S}_l\right) dT = \left(+\overline{V}_l - \overline{V}_s\right)_s dP$$

for the entropy and volume changes for the phase transition

$$\Delta S = S_l - S_s \qquad\qquad \Delta V = V_l - V_s$$
$$\Delta\overline{S} dT = \Delta\overline{V} dP$$

to give the slope of the solid–liquid line on a phase diagram.

$$dP/dT = \frac{\Delta S}{\Delta V}$$

The entropy of melting at any temperature near the normal (1 atm) melting point is

$$\Delta S_m = \frac{\Delta H_m}{T}$$

and

$$\frac{dP}{dT} = \frac{\Delta H_m}{T \Delta V_m}$$

for liquid–solid or solid–solid phase transitions.

The Clapeyron equation gives the slope of the equilibrium (2 phase) line which separates a solid from a liquid in a pressure–temperature phase diagram. The entropy change is positive on melting. An increase in volume on melting produces a positive slope. Water is an exception. The volume of water decreases on melting and the slope of the ice-liquid water transition is negative.

The slopes of the solid–liquid equilibrium lines are generally quite large because the volume change is small. An entropy change of +1 Latm mol^{-1} with a volume increase of 0.02 L mol^{-1} gives a slope

$$\frac{dP}{dT} = \frac{\Delta S}{\Delta V} = \frac{1\,\text{Latm mol}^{-1}\text{K}^{-1}}{0.01\,\text{L mol}^{-1}} = 100\,\text{atm K}^{-1}$$

A one degree change in the temperature requires a 100 atm change in the pressure to maintain the equilibrium. The line separating the liquid and solid phases in the phase diagram appears vertical.

9.9 The Clausius–Clapeyron Equation

In a pressure–temperature phase diagram, liquid–solid and solid–solid equilibrium lines are straight because the entropy change and volume changes are constant over a large temperature range. By contrast, the liquid–vapor equilibrium line curves with an exponential dependence.

When one mole of a liquid vaporizes, 1 or 2 mL of the liquid expands to roughly 20 L of volume as vapor. The volume difference is essentially the vapor volume

$$\Delta \overline{V} = \overline{V}_{vap} - \overline{V}_{liq} \approx \overline{V}_{vap}$$

Substituting into the Clapeyron equation

$$\frac{dP}{dT} = \frac{\Delta \overline{S}}{\Delta \overline{V}} = \frac{\Delta \overline{H}}{T \overline{V}_{vap}}$$

and introducing the ideal gas volume per mole

$$\overline{V} = V/n = RT/P$$

gives

$$\frac{dP}{P} = \frac{\Delta H}{RT^2} dT$$

The variables P and T are separated for integration

$$\frac{dP}{P} = \frac{\Delta H}{RT^2}dT$$

The vapor pressure of the liquid is 1 atm at its boiling point by definition. An integration from the normal boiling point at 1 atm gives the vapor pressure at T

$$\int_1^P \frac{dP}{P} = \int_{T_{bp}}^T \frac{\Delta H}{RT^2}dT \qquad \ln\frac{P}{1} = -\frac{\Delta H}{R}\left(\frac{1}{T} - \frac{1}{T_{bp}}\right)$$

The slope of a $\ln(P)$ versus $1/T$ plot

$$d\ln P = \frac{\Delta H}{R}d\left(\frac{1}{T}\right) \qquad \frac{d\ln P}{d(1/T)} = -\frac{\Delta H}{R}$$

is proportional to the enthalpy of vaporization

$$slope = -\frac{\Delta H}{R}$$

This linear plot format parallels the Gibbs–Helmholtz equation because the pressure equilibrium constant for the phase transition

$$A(\text{liq}) = A(\text{vap})$$

is $K_p = p$, the equilibrium vapor pressure.

9.10 Freezing Point Depression

If ice and water are at equilibrium and a solute is added to the water at $0°C$, the chemical potential of the water is lower than that of the ice. Ice melts completely trying to restore the ice-water equilibrium at this temperature. For equilibrium, the freezing point must decrease.

Because ice is solute free, only water equilibrates. The chemical potential balance equation

$$d\,\mu_{\text{solution}}(H_2O) = d\,\mu_{\text{solid}}(H_2O)$$
$$-\bar{S}_i dT + RT\ln X_i = -\bar{S}_s dT + RT d\ln X_w$$

has $X_{\text{ice}} = 1$ since no solute appears in ice at equilibrium,

$$-\bar{S}_s dT = -\bar{S}_w dT + RT d\ln X_w$$
$$\left(\bar{S}_w - \bar{S}_s\right) dT = \Delta\bar{S}_{\text{fus}} dT = RT d\ln X_w$$

The entropy difference for melting is

$$\Delta \bar{S}_{\text{fus}} = \frac{\Delta \bar{H}_{\text{fus}}}{T}$$

so that

$$\frac{\Delta H_{\text{fus}}}{T^2} dT = R d\ln(X_{\text{w}})$$

The integration limits are $(X_{\text{w}} = 1, T = T_{\text{f}})$ to (X_{w}, T)

$$\int_{T_{\text{f}}}^{T} \frac{\Delta H_{\text{fus}}}{T^2} dT = R \int_{1}^{X_{\text{w}}} d\ln X \qquad \frac{\Delta H_{\text{fus}}}{R}\left(\frac{1}{T_{\text{f}}} - \frac{1}{T}\right) = \ln \frac{X_{\text{w}}}{1}$$

For a dilute solution, $T \approx T_{\text{f}}$. The reciprocal temperature difference is

$$\frac{1}{T_{\text{f}}} - \frac{1}{T} = \frac{T - T_{\text{f}}}{TT_{\text{f}}} \approx -\frac{\Delta T}{T_{\text{f}}^2}$$

where

$$\Delta T = T_{\text{f}} - T$$

and

$$-\frac{\Delta H_{\text{fus}}}{RT_{\text{f}}^2} \Delta T = \ln(X_{\text{w}})$$

Since the solution is dilute,

$$X_{\text{w}} = 1 - X_{\text{s}}$$
$$\ln X_{\text{w}} = \ln(1 - X_{\text{s}}) \approx -X_{\text{s}}$$

The freezing point depression

$$-\frac{\Delta H_{\text{fus}}}{RT^2} \Delta T = -X_{\text{s}}$$

compares with the "usual" equation

$$\Delta T = K_{\text{f}} m$$

where K_{f} is a constant and m is the molality of the solution.

A 2 m solution has 2 moles of solute per 1000 g of solvent. If this solvent has a molecular weight M', there are

$$n' = 1000/M'$$

moles of solvent so the mole fraction of solute is

$$X_s = \frac{2}{\frac{1000}{M'} + 2} = \frac{m}{m + \frac{1000}{M'}}$$

$$X_s \approx \frac{m}{\frac{1000}{M'}}$$

so that

$$\frac{\Delta H_{fus}}{RT^2} \Delta T = \frac{mM'}{1000}$$

The freezing point depression

$$\Delta T = \left(\frac{M'}{1000}\right) \left(\frac{RT^2}{\Delta H_{fus}}\right) m$$

gives K_f in solvent parameters

$$K_f = \frac{M'RT^2}{1000\Delta H_{fus}}$$

The freezing point depression constant for water with a molecular weight of 18 and an enthalpy of fusion of 601.9 J g^{-1} is

$$K_f = \frac{18\,g\,mol^{-1}8.31\,J\,mol^{-1}K^{-1}273\,K^2}{1000 \times 601.9\,J\,mol^{-1}} = 1.86\,K\,m^{-1}$$

9.11 Boiling Point Elevation

The freezing point is depressed when solute is added to the liquid phase. The boiling point increases when solute is added to the liquid phase to compensate the decrease in liquid chemical potential. The chemical potential balance equation for equilibrium

$$d\mu_{liq} = d\mu_{vap}$$

$$-\bar{S}_{solv}dT + RTd\ln X_{solv} = -\bar{S}_{vap}dT$$

is integrated from (X_1=1, T_{bp}) to (X_1, T) Using

$$\Delta\bar{S} = \frac{\Delta\bar{H}}{T}$$

gives

$$\int_{T_b}^{T} -\frac{\overline{H}_v}{T^2}dT + \int_{1}^{X} R\, d\ln X = \int_{T_b}^{T} -\frac{\overline{H}_v}{T^2}dT$$

$$-\frac{\Delta H_v}{R}\left(\frac{1}{T} - \frac{1}{T_b}\right) = -\ln X_{solv} = +X_s$$

The reciprocal temperature difference becomes

$$\left(\frac{1}{T} - \frac{1}{T_b}\right) = -\frac{\Delta T}{T_b^2}$$

where ΔT is now the positive temperature difference. The final expression is

$$\frac{\Delta H}{R}\frac{\Delta T}{T_b^2} = X_s$$

The mole fraction is converted to molality

$$\frac{\Delta H_v}{R}\frac{\Delta T}{T_b^2} = \frac{mM'}{1000}$$

and a boiling point elevation

$$\Delta T = \frac{M'RT^2}{1000\Delta H_v}m$$

9.12 Donnan Equilibrium

Dissolved salts that dissociate in solution are electroneutral. A Donnan equilibrium occurs when two ionic solutions are separated by a membrane permeable to specific anions and cations. The equilibrium also occurs when one phase with fixed charge, e.g., charged proteins or ion exchange beads, is dispersed in water.

For KCl, the total free energy of each phase is the sum of the chemical potential of each ion

$$1d\,\mu_1(K) + 1d\,\mu_1(Cl) = 1d\,\mu_2(K) + 1d\,\mu_2(Cl)$$
$$1RTd\ln[K]_1 + 1RTd\ln[Cl]_1 = 1RTd\ln[K]_2 + 1RTd\ln[Cl]_2$$

gives

$$[K]_1\,[Cl]_1 = [K]_2\,[Cl]_2 = K_{sp}$$

The equilibrium balances products of concentrations of the anions and cations (a solubility product):

$$K_{sp} = [K^+][Cl^-]$$

For $CaCl_2$, the free energy balance equation requires 1 mole of Ca^{2+} and 2 moles of Cl^-. The balance equation is

$$1RTd\ln[Ca]_1 + 2RTd\ln[Cl]_1 = 1RTd\ln[Ca]_2 + 2RTd\ln[Cl]_2$$

and

$$[Ca]_1[Cl]_1^2 = [Ca]_2[Cl]_2^2$$

The Donnan equilibrium explains the experimental observation of salt gradients across membranes or phases with no energy source to produce them. The concentration of an anion or cation is increased by adding a second salt with that ion and a counterion that is impermeable or locked as fixed charge in one phase. For example, all salts that release independent K^+ in solution contribute to the total K^+ concentration in the phase.

Two solutions are separated by a membrane permeable to both K^+ and Cl^-. A univalent potassium protein salt (K^+P^-) with membrane impermeable P^-, is now added to bath 1 to increase the total K^+ concentration in solution. KCl moves from bath 1 to bath 2 to maintain the equilibrium, i.e., keep the KCl concentration products in both solutions equal. This reduces the KCl concentration in bath 1.

A non-equilibrium system which has 1 mM KCl and 1 mM KP in bath 1 and 1 mM KCl in bath 2 moves toward an equilibrium. The product of K^+ and Cl^- is larger on side 1 because the K^+ concentration is 2 mM, KCl flows to bath 2 to reestablish the equilibrium. If x mM K^+ flows from bath 1 to bath 2, it must be accompanied by x mM of Cl^- to maintain electroneutrality. The 2 mM K^+ in bath 1 (1 mM from KCl + 1 mM from KP) decreases by x at equilibrium. The 1 mM chloride concentration drops by x as well to maintain the electroneutrality of bath 1. Bath 2 contains only 1 M KCl so both the K^+ and Cl^- concentrations increase by x.

The Donnan equilibrium for KCl between the two phases is now

$$[K]_1[Cl]_1 = [K]_2[Cl]_2$$
$$(2 - x)(1 - x) = (1 + x)(1 + x)$$
$$2 - 3x + x^2 = 1 + 2x + x^2$$
$$5x = 1$$
$$X = 0.2$$

x is substituted into the expressions for each species to determine their equilibrium concentrations,

$$[K]_1 = 2 - x = 1.8, \qquad [Cl]_1 = 1 - x = 0.8$$
$$[K]_2 = [Cl]_2 = 1 + x = 1.2$$

Each phase is electrically neutral. Bath 1 has 1.8 mM of positive charge which is counterbalanced by 0.8 mM of Cl^- and 1 mM of the P^- protein that induced the flow to bath 2. Bath 2 has 1.2 mM of both K^+ and Cl^-.

The new concentrations satisfy the Donnan equilibrium condition,

$$[K]_1[Cl]_1 = [K]_2[Cl]_2$$

1:1 electrolytes are very tractable because the quadratic terms cancel. In salts such as $CaCl_2$, the resulting equation includes a third power in x. In addition, electroneutrality requires that 2 Cl^- ions travel with each Ca^{2+}. The Donnan equilibrium expression for 1 mM $CaCl_2$ and 1 mM CaP in bath 1 is

$$[Ca]_1[Cl]_1^2 = [Ca]_2[Cl]_2^2$$

For KCl, the concentrations of both K^+ and Cl^- differ in each of the phases. If K^+ sensitive electrodes were inserted into each bath, a Donnan electrical potential difference could be recorded as

$$\Delta\psi = -\frac{RT}{+1F}\ln\frac{1.2}{1.8} = 25\ln\left(\frac{2}{3}\right) = -10.14\,\text{mV}$$

On the other hand, if a Cl^- ion-sensitive electrode were used, the measured potential difference would be

$$\Delta\psi = -\frac{RT}{-1F}\ln\frac{[Cl]_2}{[Cl]_1} = +25\ln\frac{1.2}{0.8} = 25\ln\frac{3}{2} = +10.14\,\text{mV}$$

The magnitude of the potential is the same; only the sign of the potential changes. In order to measure a potential, the electrode must be sensitive to only one of the two equilibrated species.

Problems

9.1 A fiber, e.g., muscle, produces a force τ when placed in an ionic solution of molar concentration c. For a length L of fiber, the chemical potential is $-Ld\tau$. The force decreases as the concentration increases.

a. Define a chemical potential for this system using ion concentration c and the force τ as independent variables.
b. If one portion of fiber is placed in a bath of length L with a concentration c_1 of ions and second portion is placed in a second bath of length L with a concentration c_2 of ions, the two portions of the fiber will come to equilibrium at different lengths. Give the balance equation.

9.2 Two metals with absolute entropies S_1 and S_2, respectively, are placed together. If a potential difference (ψ_1 for metal 1 and ψ_2 for metal 2) is applied to the pair, the temperatures of the two metals will differ at equilibrium.

a. Set up the appropriate chemical potential balance equation.
b. Use the enthalpies for each phase (H_1 and H_2) to modify your equation in (a) so that it depends on the enthalpy difference between the phases.
c. Determine the natural logarithm of the ratio of temperatures produced by applying a potential difference $\Delta\psi = \psi_2 - \psi_1$ between the two metals. Both metals have the same temperature when the potential difference is 0.

9.3 The K^+ salt of a protein of 50,000 daltons gives an osmotic pressure of 0.001 atm when 0.05 g is dissolved in H_2O to make 0.1 L of solution. $RT = 25$ Latm.

a. What is the experimental molecular weight of this protein from the osmotic pressure experiment?
b. How many K^+ ions are present per protein molecule?

9.4 A membrane separates two NaCl solutions of equal concentration (0.1 M) and is permeable to both ions. Determine the concentration of a second salt NaA with impermeable anion A^- that must be added to bath 1 to transfer 0.01 M NaCl from solution 1 to solution 2.

9.5 At 300 K and 1 atm, $\Delta G = 75$ J mol^{-1} for the conversion of monoclinic sulfur ($V = 16.3 \times 10^{-3}$) to rhombic sulfur($V = 15.5 \times 10^{-3}$) Estimate the minimum pressure necessary to achieve a stable monoclinic phase at this temperature.

9.6 A fritted disk is used to separate mercury liquid and mercury vapor. If the pressure on the liquid is increased from 1 to 11 atm, what is the change in vapor pressure of the gaseous mercury. The initial vapor pressure is 1 atm and the partial molar volume of the liquid is 2.46 L mol^{-1}.

9.7 Two solutions with concentrations of 0.1 M are separated by a semipermeable membrane. An additional 0.5 atm of pressure is now applied to bath 1 and the system is allowed to equilibrate.

a. Determine the ratio of mole fractions for equilibrium with the additional pressure.

9.8 At equilibrium, an aqueous phase (2) has a KCl concentration of c_2 and a protein phase (1) has a KCl concentration of c_1 and a K_zP^{z-} concentration of c_p.

a. Show

$$[c_1 + zc_p][c_1] = c_2^2$$

9.9 A semipermeable bag with 1 L of a sugar solution is placed in pure water and the bag swells to an equilibrium volume of 11 L.

a. Set up a potential balance equation in the concentrations and the surface energy of the bag.
b. If the bag energy per unit surface area is 12 J L^{-1} determine the water mole fraction of the sugar solution.

9.10 An n-type semiconductor is doped with atoms that provide electrons that can move to a second semiconductor where the concentration of such electrons is lower. These electrons carry a negative charge which moves with the electron to the lower concentration region to produce an electrical potential difference.

a. Set up a chemical potential balance equation involving the electrons concentrations (c_h and c_l) and the potentials (ψ_H and ψ_L).
b. Starting from the equilibrium condition $c_H = c_L = c_0$, with equal electrical potentials, determine the potential difference that arises when the electrons reach their equilibrium concentrations (c_H and c_L).

Chapter 10
The Foundations of Statistical Thermodynamics

10.1 The Ergodic Hypothesis

Classical thermodynamics does not require detailed molecular information for thermodynamics parameters. The heat capacity is a proportionality constant relating E to T:

$$E = C_v T$$

Statistical thermodynamics uses microscopic parameters such as molecular energies and probability to generate averages. The average energy, for example, matches the energy determined via classical thermodynamics.

A single system with an Avogadro's number of particles gives stable averages. Even though energy is transferred continually between particles on collision, the number with each energy remains constant for constant temperature and volume.

The averaged parameters are obtained as either time or number averages. In the first case, a single system is examined at a series of different times and parameters observed at each time are averaged.

In the second case, the parameters for a large number of identical systems, the ensemble, are observed "simultaneously." The parameter values observed for each separate system are then averaged. The accuracy of the ensemble average improves as the number of systems in the ensemble increases.

A system of 100 particles, each of which has two possible energies ε_1 or ε_2 is ensemble averaged by examining four such identical systems. The first system has 74 particles with energy ε_1 and 26 with energy ε_2 (74, 26), the second has (76, 24) and the third and fourth ensemble systems have (75, 25). The ensemble average is then a system which has 75 particles with energy ε_1 and 25 particles with energy ε_2.

The time average for a single system might show the particles with the two allowed energies in the combination (75, 25). A second observation at a later time might be (77, 23), a third (74, 26), and a fourth (74, 26). This time average also gives an average distribution having 75 particles with energy ε_1 and 25 particles with energy ε_2.

The ergodic hypothesis states that time and ensemble averages are equal. This hypothesis might seem intuitively obvious but is difficult to prove except for special cases.

M.E. Starzak, *Energy and Entropy*, DOI 10.1007/978-0-387-77823-5_10,
© Springer Science+Business Media, LLC 2010

The variation in the time average reflects the fact that energy is continuously transferred between the molecules of the system. A state of the system would involve each of the 100 labeled molecules with a specific energy. For example, molecules 1 through 75 might have energy ϵ_1 while those labeled 76–100 have energy ϵ_2. Another state could have 1–74 and 76 with energy ϵ_1. 75 and 77–100 then have energy ϵ_2. There are clearly a large number of states that can give the result (75, 25) which is called a distribution. An observer sees the distribution (75, 25) which could be any of the labeled states that have 75 particles with energy ϵ_1.

A system observed in time must ultimately return to its initial state. If all states are equally probable, the system will, on average, pass through every possible state in the system before returning to the initial state. The number of states is then equal to the number of events or observations required before regaining the initial state. This sequence is a Poincare cycle. If the time per event is known, the total time to return to the initial state is the time per state multiplied by the total states of the system, the Poincare recurrence time.

10.2 States and Distributions

An ideal gas, expanding spontaneously from 1 to 2 L increases its entropy because the number of locational states doubles. All such states are equally probable including the one with all the particles in the original 1 L. However, for large numbers of particles, this original state, while possible, is never observed in practice. The expanded gas remains spread homogeneously in the 2 L container. Even situations where all but 2 or 3 particles appear on the left are not seen. For large numbers of particles, 50% of the particles are found in the left container. A state with 100%, while just as probable as a 50–50 state is "almost" never observed.

The observation that the 2 L container has equal particles in each liter seems to contradict the postulate, proposed by Boltzmann, that all states are equally probable. A state with all particles in the left container is just as likely as one with the particles equally divided between left and right. In reality, there is no contradiction. The observer observes only the distribution of particles without the labels that define a specific state. Many states will produce a 50–50 distribution, while only a single state will produce the 100–0 distribution (all particles in the left 1 L). The distribution is the number of particles in each container or the fraction of particles in each bulb.

$$\left[p\,(\text{left}), p\,(\text{right})\right] = [0.5, 0.5]$$

States describe the way labeled particles are distributed between the two 1 L containers. A 50–50 distribution includes all the states with half the particles on the left. For 100 particles, there are 100 states with the [99, 1] distribution – particle 1 left, particle 2 left, . . ., particle 100 left).

A 50–50 distribution of particles between the two halves of a container at equilibrium has the largest number of distinct states. This is the equilibrium distribution. Other distributions can appear as fluctuations but the 50–50 dominates because it has the largest number of the equally probable states.

10.3 The Dog-Flea Model

The distinction between states and distributions was difficult to understand because Boltzmann developed it for large numbers of particles. His student, Paul Ehrenfest, developed heuristic models to illustrate the differences with simple systems.

Although Ehrenfest ultimately published his work as the "urn" model, it was also described as a "dog-flea" model. The two urns (or 1 L containers) are dogs. A fixed number of fleas (particles) are free to jump between the two dogs. This is equivalent to particles moving between the two connected 1 L containers or balls moved between two urns. The fleas jump between the dogs randomly so an initial state, e.g. (100, 0), becomes a new state when a flea jumps.

The model is also identical to a chemical reaction with two isomers A and B of equal energy. Either isomer can randomly convert to the other in some time interval.

A system with two fleas and two dogs has four states.

(1) Fleas 1 and 2 on dog A.
(2) Fleas 1 and 2 on dog B
(3) Flea 1 on dog A, flea 2 on dog B
(4) Flea 2 on dog A, flea 1 on dog B

Equivalently, for isomers (A;B)

(1)   (1,2;0)
(2)   (0;1,2)
(3)   (1;2)
(4)   (2;1)

Each of the four states is equally probable. Two of the four states have a 50–50 distribution and this distribution occurs twice as often as the (2;0) or (0;2) distribution. As the number of fleas increases, this equilibrium distribution encompasses an increasingly larger fraction of the total states. For a mole of fleas or isomers, this equilibrium 50–50 distribution is characterized by a dominant fraction of the states. Most of the remaining states are found in distributions similar to the 50–50 distribution, e.g., (49, 51) or (51, 49).

The relative increase in both total states and states for a 50–50 distribution is illustrated with four fleas on two dogs, i.e., four molecules with two isomeric forms. Since each molecule is either A or B (two possibilities), the total number of states possible is

$$2^4 = 16$$

The number of states in the 50–50 distribution (2, 2) is

$$\frac{4!}{2!2!} = 6$$

The 4! Is the number of possible orders for the four labeled isomers. Since the order of labeled particles for each isomer is irrelevant, e.g. (1,2;34) is the same state as (2,1;3,4), the ways of ordering the A isomers(or dog A) reduces the total states by a factor of 2!. The ordering of B also reduces the total number of allowed states by 2!. By contrast, 4 fleas on dog 1 appear in 4! Orders and the number of states is

$$4!/4!0! = 1$$

The (4, 0) distribution has a probability of 1/16.

The 2 flea–2 dog model had a higher 50–50 probability (0.5) than the 2 dog–4 flea model (0.375) but the overall distribution is narrowing. Only $2/16 = 1/8$ of the states are (4, 0) or (0, 4).

For only 50 particles with two isomers (far less than an Avogadro's number of particles), the total number of states is

$$2^{50} = 1.1259 \times 10^{15}$$

Of this total, the number of states that give 25 particles of each isomer is

$$\frac{50!}{25!25!} = 1.264 \times 10^{14}$$

The probability of finding this distribution is

$$\frac{1.264 \times 10^{14}}{1.1259 \times 10^{15}} = 0.112$$

The (24, 26), (26, 24), (27, 23), etc. distributions also have high probabilities. The range of distributions with significant probability is narrowing but not yet spiking at 50–50. The fraction of states for only A isomers or all fleas on dog 1,

$$\frac{1}{1.1259 \times 10^{15}} = 8.8 \times 10^{-16}$$

is extremely small indicating the improbability of observing this distribution.

An isomerization reaction passes through all the states in time. A system starting with all A isomers must ultimately return to this state as the system cycles through all the states of the system. This is a Poincare cycle. If each state change takes time t, the time for a Poincare cycle is Ωt. A two molecule system has $2^2 = 4$ states and the Poincare cycle time is $4t$.

For 100 molecules, probability of the (100, 0) distribution is

$$p = \frac{1}{2^{100}} = 7.9 \times 10^{-31}$$

$\Omega = 1.2 \times 10^{30}$ and the Poincare cycle time is $1.2\Omega 10^{30}t$. Even molecules with collisions every 10^{-25} s require
$1.2 \times 10^{30} \times 10^{-25} = 1.2 \times 10^5$ s = 2 days. With an Avogadro's number of particles, the time increases dramatically. The chance of returning to the original state is finite but ridiculously small.

10.4 The Most Probable Distribution

The equilibrium distribution for two isomers of equal free energy contains the largest number of states. The systems with 2 and 4 isomers had binomial distributions of states (1:2:1) and (1:4:6:4:1). The distributions are symmetric with a maximum at the most probable distribution 50–50 distribution. With increasing particles, the binomial distribution becomes a symmetric Gaussian distribution with a maximum at the 50–50 distribution.

The number of states for each distribution is encased in the binomial distribution for N particles and 2 isomers

$$(x + y)^N = \sum_{n=0}^{N} \frac{N!}{n!(N - n)!} x^n y^{N-n}$$

For a total of N particles, A distribution with n_A A isomers and $n_B = N - n_A$ B isomers has of states

$$\Omega = \frac{N!}{n_A! n_B!}$$

states

$$N = n_A + n_B$$

The most probable distribution and average distribution have

$$n_A = n_B = N/2$$

The most probable distribution is obtained by differentiation with respect to n_A or n_B. Since

$$n_B = N - n_A$$

the binomial coefficient is

$$\Omega = \frac{N!}{n_A! (N - n_A)}$$

The factorials cannot be differentiated but their logarithms can. The logarithm of this function has the same maximum. The factorials are converted to differentiable functions using Stirling's approximation,

$$\ln M! = M \ln M - M$$

The derivatives for $\ln n_A!$ and $\ln(N - n_A)!$ Are

$$d (\ln n_A!) / dn_A = d [n_A \ln (n_A) - n_A] / dn_A$$
$$= \ln (n_A) + n_A / n_A - 1 = \ln (n_A)$$
$$d \ln (N - n_A)! / dn_A = d [(N - n_A) \ln (N - n_A) - (N - n_A)] / dn_A$$
$$= - \ln (N - n_A) - (N - n_A) / (N - n_A) + 1 = - \ln (N - n_A)$$

The logarithm of the binomial

$$\ln \Omega = \ln \frac{N!}{n_A! (N - n_A)!}$$

becomes

$$\ln \Omega = \ln N! - \ln n_A! - \ln (N - n_A!) = N \ln N - (n_A)$$
$$\ln n_A + n_A - (N - n_A) \ln (N - n_A) + (N - n_A)$$

Differentiating with respect to n_A and equating to zero

$$\{d \ln \Sigma\} / \{dn_A\} = 0 = - [\ln (n_A) + \ln (N - n_A)] = 0$$

$$\ln n_A = \ln (N - n_A)$$
$$n_A = N - n_A$$
$$\#2n_A = N$$
$$n_A = n_B = \frac{N}{2}$$

As expected, a 50–50 distribution ($n_A = N/2$) is most probable with the maximal number of states.

10.5   Undetermined Multipliers

A single derivative suffices to determine the most probable distribution for the two isomer model since n_B is expressed in terms of n_A. For a system with three or more equal energy isomers, the binomial equation cannot be reduced to a single differentiable variable. The method of undetermined multipliers resolves this difficulty. For two isomers and N particles, the constraint

$$N = n_A + n_B$$

is rearranged

$$N - n_A - n_B = 0$$

multiplied by the undetermined coefficient α and added to the general expression for the number of states

$$\ln\{N!/\,(n_A!)\,(n_B!)\} + \alpha\,(N - n_A - n_B)$$

α is determined using the constraint. The constraint term equals zero but the derivative is finite.

Since

$$d\,(\ln(n_A!))\,/dn_A = \ln(n_A)$$

The derivative selects only terms in n_A

$$0 = -\ln n_A + \alpha(-1)$$
$$n_A = \exp(-\alpha)$$

Differentiation with respect to n_B gives

$$n_B = e^{-\alpha}$$

Since α is the same for both equations, $n_A + n_B = e^{\alpha} + e^{\alpha} = 2e^{\alpha}$

$$e^{\alpha} = n_A = n_B = \frac{N}{2}$$

The approach work for multiple (p) isomers

$$\ln \frac{N!}{n_1!n_2!\ldots n_p!} + \alpha\,(N - n_1 - n_2 - \ldots - n_p)$$

Each n_i is maximized

$$\frac{\partial}{\partial n_i}\left[\ln \frac{N!}{n_1!n_2!\ldots n_i!\ldots n_p!} + \alpha\,(N - n_1 - n_2 - \ldots - n_i\ldots - n_p)\right] = -\ln n_i - \alpha$$

$$n_i = e^{\alpha}$$

$$N = \sum_1^p e^{-\alpha} = pe^{\alpha}$$

$$n_i = N/p$$

The populations of each of the isomers are the same.

10.6 Energy Distributions

The isomers (or fleas) that form a distribution all had exactly the same energy. The situation becomes more realistic and more complicated if the system consists of a molecule with different energies. The distribution is the number of molecules with energy ε_0, molecules with energy ε_1, etc. Because the total amount of energy is limited, the distribution need not have equal numbers of molecules at each energy. In fact, high-energy molecules appear less frequently than low-energy molecules in such distributions.

The isomerization

$$A \rightleftharpoons B$$

has reference energy 0 for A and free energy ΔG for B. The equilibrium distribution is not 50–50; the high-energy species appears less frequently at equilibrium.

Systems where each particle, isomer or flea have exactly the same energy are microcanonical ensembles. Formally, the ensemble is a large number of systems containing a fixed number of particles that can distribute in two (or more) equi-energy ways. The average distribution is then the average of all these independent ensembles.

The molecule with, for example, three possible energies is analogous to a system with three different isomers. The system is a mixture of molecules with each of the energies. Within the constraint of limited total energy, the equilibrium distribution subsumes the largest number of states.

A system where the molecules are distinguished by their different energies is a canonical ensemble. Again, the ensemble is a large number of identical systems that have the same fixed number of molecules and equal total energy.

The canonical ensemble has two constraints that become undetermined multipliers. The system has a fixed number of molecules. For N molecules

$$N = n_0 + n_1 + \ldots + n_i + \ldots = \sum_{i=0}^{\infty} n_i$$

n_0 of the molecules have energy ε_0, n_1 have energy ε_1, etc. The total energy of all ε_0 molecules is

$$\varepsilon_0 n_0$$

The n_1 molecules with energy ε_1 each contribute

$$\varepsilon_1 n_1$$

The total energy E for the system is

$$E = n_0 \varepsilon_0 + n_1 \varepsilon_1 + \ldots + n_i \varepsilon_i + \ldots = \sum_{i=0}^{\infty} n_i \varepsilon_i$$

These two constraints establish the equilibrium distribution for the canonical ensemble.

The total number of states possible with n_0, n_1, etc. molecules is

$$\Omega = \frac{N!}{n_0! n_1! \ldots n_i! \ldots} = \frac{N!}{\overset{\infty}{\underset{i=0}{\pi}} n_i}$$

The two undetermined multiplier constraints are added

$$\ln \frac{N!}{\overset{\infty}{\underset{i=0}{P}} n_i} + \alpha \left(N - \sum_{i=0}^{\infty} n_i \right) + \beta \left(E - \sum_{i=0}^{\infty} n_i \varepsilon_i \right)$$

and differentiated with respect to each n_i for the equilibrium distribution with maximal states. Only three terms survive differentiation with respect to each n_i

$$\frac{d}{dn_i} [-\ln n_i! - \alpha n_i - \beta n_i \varepsilon_i] = 0$$

$$\frac{d}{dn_i} [-\ln n_i! - \alpha n_i - \beta n_i \varepsilon_i] = -\ln n_i - \alpha - \beta \varepsilon_i = 0$$

$$n_i = e^{-\alpha} e^{-\beta \varepsilon_i}$$

The sum of all the n_i must equal N

$$N = \sum_{i=0}^{\infty} n_i = \sum_{i=0}^{\infty} e^{-\alpha} e^{-\beta \varepsilon_i}$$

$\exp(-\alpha)$ is common to all terms. Factoring and solving

$$e^{-\alpha} = \frac{N}{\sum_{i=0}^{\infty} e^{-\beta \varepsilon_i}}$$

Substituting for $\exp(-\alpha)$

$$n_i = N \frac{e^{-\beta \varepsilon_i}}{\sum_{i=0}^{\infty} e^{-\beta \varepsilon_i}}$$

The probability of finding a particle with energy ε_i is

$$p_i = \frac{n_i}{N} = \frac{e^{-\beta \varepsilon_i}}{\sum e^{-\beta \varepsilon_i}}$$

The sum of all exponentials (the Boltzmann factors) in the denominator converts each Boltzmann factor (a number) into a probability. This sum over Boltzmann factors is called a "sum over states" or partition function. Neither definition accurately reflects this "sum over Boltzmann factors in each possible energy" but partition function has the widest usage.

Energy isomers are not equally probable. The exponential value decreases with higher energies. The number of states with higher energy is always smaller. Even though the state populations differ, the Boltzmann distribution gives the largest number of states consistent with the constraints.

10.7 The Boltzmann Factor

The probabilities p_i for the canonical system contain the second undetermined multiplier. For microscopic (single molecule) energies

$$\beta = \frac{1}{kT}$$

This result is inferred from the following observations:

(1) kT is positive so the exponential is smaller for higher energies;
(2) kT has units of energy per molecule; the exponential ratios are dimensionless; and
(3) the energy is proportional to T.
(4) The exponential has appeared for thermodynamic systems such as the barometric formula:

$$\frac{p}{p_o} = e^{\frac{mgh}{kT}}$$

and Gibbs–Helmholtz equation (h is the enthalpy per molecule)

$$p \propto \exp\left(-h/kT\right)$$

$\beta = 1/kT$ is established rigorously in Chapter 11.

10.8 Bose–Einstein Statistics

Two dogs that contain three fleas have a total of $2^3 = 8$ possible states. The (3, 0) and (0, 3) states don't require labeled fleas. The distribution (2, 1) has three states in Boltzmann statistics (– flea 1, 2, or 3 on dog 2), with labels [D1;D2] = [1,2;3], [1,3;2], and [2,3;1].

Boltzmann statistics recognizes which labeled flea appears on Dog 2. Bose–Einstein statistics, by contrast, uses indistinguishable particles. For Boltzmann statistics, each of the three Boltzmann states is a single state with 2 indistinguishable particles on dog 1 and one indistinguishable particle on dog 2.

For two dogs and three indistinguishable fleas, Bose–Einstein statistics predicts
[3, 0],[2, 1],[1, 2], [0, 3]
For a total of four states rather than the $2^3 = 8$ total states of Boltzmann statistics.

The four Bose–Einstein states are also expressed using x for each indistinguishable flea and "|" to separate the dogs

$$(xxx|) \quad (xx|x) \quad (x|xx) \quad (|xxx)$$

The indistinguishable particles can be discrete quanta occupying a single vibration in a molecule. The 3-flea/2-dog model becomes a molecule with two identical quantized vibrations. The three photons then distribute into these two vibrational modes following Bose Einstein statistics to give the four states.

The number of Bose–Einstein states in this case arranges the quanta (x) and partitions (|) along a line to determine all possible arrangements. The four Bose–Einstein states are

$$Xxx| \quad xx|x \quad x|xx \quad |xxx$$

There are 4! ways of arranging these four entities. The order of the x's on the line must be factored out by dividing by 3!. Only $2-1 = 1$ "1" entities are required to create the two separate vibrational modes. The factor (1!)[not 2!] must be factored

$$\frac{4!}{3!1!} = 4$$

The equation for the number of states when there are n quanta in g identical oscillators is

$$\Omega = \frac{(n + g - 1)!}{n!(g - 1)!}$$

This formula replaces the g^n total states for Boltzmann statistics and is useful when there are a relatively small number of oscillators and quanta. The total states are used to determine the distribution of energy among vibrational modes in a single molecule. For systems where both g and n are large, the n and g for the maximal states are found using undetermined multipliers and the constraints

$$N = \sum ni$$

$$E = \sum n_i \varepsilon_I = \sum n_i i\varepsilon$$

where ε is the energy of a single quantum. The n_i for the distribution with the largest number of states is

$$\frac{\partial}{\partial n_i} \left[\ln \left(\frac{(n_i + g - 1)!}{n_i!(g - 1)!} \right) + \alpha (N - \Sigma n_i) + \beta (E - \Sigma n_i \varepsilon_i) \right] = 0$$

$$\ln (n_i + g - 1) - \ln (n_i) - \alpha - \beta \varepsilon_i = 0$$

$$\frac{n_i}{n_i + g - 1} = e^{-\alpha} e^{-\beta \varepsilon_i}$$

The n_i in this equation is that for the distribution giving the maximal number of states subject to the two constraints. It is useful when g and n_i are large. In this case $g-1 \approx g$. The equation is solved for n_i

$$n_i = (n_i + g)e^{-\alpha}e^{-\beta\varepsilon_i}$$

$$n_i = \frac{ge^{-\alpha}e^{-\beta\varepsilon_i}}{1 - e^{-\alpha}e^{-\beta\varepsilon_i}} = \frac{g}{e^{\alpha}e^{\beta\varepsilon_i} - 1}$$

g is the degeneracy of each of the energy states. If g can vary with energy level, the degeneracy becomes g_i and

$$\frac{n_i}{g_i} = [e^{\alpha}e^{\beta\varepsilon_i} - 1]^{-1}$$

The constant α in Boltzmann statistics is determined from the constraint

$$N = \Sigma n_i$$

Evaluation of α for the Bose–Einstein statistics is more complicated. Since these statistics often used when both N and g are small, it is then convenient to use the total number of states. A molecule containing g equivalent vibrational modes that can contain a total of N quanta (bosons) has

$$\frac{(N + g - 1)!}{N!(g - 1)!}$$

total states.

10.9 Fermi Dirac Statistics

The dog-flea model considered the situation where a large number of fleas (n) were present on a small number of dogs (g) and the number of states was based on the number of arrangements for the N fleas. For n_1 fleas on dog 1,

$$\Omega = \frac{N!}{n_1!\,(N - n_1)!}$$

Fermi Dirac statistics are used when the number of dogs is large compared to the number of fleas. For example, each dog might have no more than one flea. In this case, the number of states is determined by the number of dogs that have a flea and the number that don't

$$\Omega = \frac{g!}{(g - n)!n!}$$

Fermi Dirac statistics are required for electrons (fermions with spin $\frac{1}{2}$) since electrons with the same quantum number can't occupy the same orbitals. This is the

Pauli exclusion principle. g is now the number of orbitals. Each electron energy level can have two orbitals (electron spins $+1/2$ and $-1/2$), i.e., $g=2$. Both can be occupied ($n=2$), one can be occupied ($n=1$) or the orbital can be empty. The number of states is repeated for each energy level so that the total number of states is a product

$$\Omega_1\Omega_2..\Omega_e\ldots = \frac{g!}{(g-n_1)!n_1!}\frac{g!}{(g-n_2)!n_2!}\cdots\frac{g!}{(g-n_i)!n_i!}\cdots$$

The logarithm of the states can be differentiated with respect to n_i to determine the value that gives the maximal states subject to the constraints

$$\sum n_i = N$$

$$\sum n_i,\varepsilon_I = E$$

For the ith level, (1 particle maximum occupancy) the only term in the logarithmic sum subject to differentiation with n_i is

$$\ln\frac{g!}{(g-n_i)!n_i!}$$

and

$$\frac{\partial}{\partial n_i}\left[\ln\frac{g!}{(g-n_i)!n!_i}+\alpha\,(N-\Sigma n_i)+\beta\,(E-\Sigma n_i\varepsilon_i)\right]=0$$

$$+\ln\,(g-n_i)-\ln\,(n_i)-\alpha-\beta\varepsilon_i=0$$

$$\ln\frac{n_i}{g-n}=-\alpha-\beta\varepsilon_i\qquad\frac{n_i}{g-n_i}=\exp\,(-\alpha)\exp\,)-\beta\varepsilon_i\,)$$

rearranging to

$$n_i=\frac{g\exp\,(-\alpha)\exp\,(-\beta\varepsilon_i)}{1+\exp\,(-\alpha)\exp\,(-\beta\varepsilon_i)}=\frac{g}{\exp\,(\alpha)\exp\,(+\beta\varepsilon_i)+1}$$

$\beta=1/kT$ and the constant α is determined since, at $T=0$ K, the energy levels will be filled continuously from the lowest level with $n=1$ to some level m with energy ε_m. $n_i=1$ for each level below m and 0 for every level above m. This is possible if $\alpha=-\beta\varepsilon_m$. At the threshold with $T=0$ K,

$$\exp\,(-\beta\varepsilon_m)\exp\,(\beta\varepsilon_m)=1$$

and level m is filled

$$n=2/(1+1)=1$$

For smaller ε ($i<m$), the exponential argument is positive and infinite when $T=0$ K. The exponential in the denominator with argument $\varepsilon-\varepsilon_m>0$ becomes vanishingly small and $n=g$. For $\varepsilon>\varepsilon_m$, the exponential is very large and $n=0$.

The analysis reflects that the levels begin to "fill" with electrons starting with lowest energy and continuing until all the electrons are added at some energy ε_m. Higher energy levels are unoccupied at 0 K. As T increases, some electrons from the highest occupied levels move into the unoccupied levels to produce a range of state populations with energy rather than the abrupt change from populated to unpopulated states observed at absolute zero.

10.10 Other Ensembles

The canonical ensemble has a fixed number of molecules per system and a constant total energy. Formally, the systems of the ensemble are in thermal contact and energy moves across the boundaries to establish the distribution populations.

The grand canonical ensemble allows both particles and energy to move between communicating systems in the ensemble, while the total energy and particles are constant. Particle transfer between systems requires an additional constraint. Just as the total energy sums products

$$n_i \varepsilon_i$$

particle transfer uses the chemical potential

$$\mu_i n_i$$

Since

$$\mu \, \alpha \, kT \ln c$$

$$c \, \alpha \, e^{\mu/kT} = \lambda$$

A grand canonical system has probabilities with Boltzmann factors for both energy and chemical potential

$$\exp(-\beta\varepsilon)\exp(+\beta\mu)$$

The full partition function must include exponentials with all possible energies and all possible molecules, i.e., a double summation

$$f_{i,N} = \frac{e^{\frac{E_i}{kT}} e^{\frac{N\mu}{kT}}}{\sum\limits_{N} \sum\limits_{i} e^{\frac{E_i}{kT}} e^{\frac{\mu N}{kT}}}$$

This formidable equation is simplified when the system is an enzyme with binding sites. The particle is then a substrate molecule and it is convenient to assume a single energy (a binding energy). In this case, the partition function for a protein with a single binding site is

$$1 + \exp(\beta\mu)\exp(-\beta\varepsilon)$$

The isobaric–isothermal ensemble permits energy flow via volume changes

$$f_{i,v} = \frac{e^{\frac{E_i}{kT}}e^{\frac{pv}{kT}}}{\sum_v \sum_i e^{\frac{E_i}{kT}}e^{\frac{pv}{kT}}}$$

where p is a single molecule pressure obtained from the macroscopic pressure by dividing by Avogadro's number.

Different ensembles have different energies but, in general, each of these energies is incorporated into the probability as a Boltzmann factor. The partition function then sums over all possible energies.

Problems

10.1 Five molecules can exist as either A or B isomers. The isomers have equal energies.

Determine the fraction of time the system has a distribution of 2 A and 3 B isomers if all states are equally probable.

10.2 Determine the distribution with maximal for a system having three equal energy isomers (1,2,3) using the undetermined multiplier technique and the constraint for the total isomers, $N = n_1 + n_2 + n_3$.

Chapter 11
Applied Boltzmann Statistics

11.1 Boltzmann Statistics for Two-Energy Levels

Systems with only two or three possible energies illustrate the Boltzmann distribution. The probabilities for such systems determine average thermodynamic parameters. The intimate connection between the energies of individual molecules and the macroscopic thermodynamic parameters makes statistical thermodynamics a powerful analytical technique.

An unpaired electron in a molecule like NO has two possible electron spins with quantum numbers $+1/2$ or $-1/2$ with the same energies, $\varepsilon_,$. At constant T, each spin state ($+1/2$ or $-1/2$) has the same Boltzmann factor

$$e^{-\frac{\varepsilon'}{kT}}$$

Since only two states are possible, the partition function sums their Boltzmann factors

$$q = e^{-\frac{\varepsilon}{kT}} + e^{-\frac{\varepsilon}{kT}} = 2e^{-\frac{\varepsilon}{kT}}$$

The probability for an electron with spin $+1/2$ is its Boltzmann factor divided by a two-term partition function

$$p_+ = \frac{e^{-\frac{\varepsilon}{kT}}}{2e^{-\frac{\varepsilon}{kT}}} = \frac{1}{2} = 0.5$$

The electron with $+1/2$ spin is found 50% of the time. The probability for the electron with -1/2 spin is also 0.5

$$p_- = \frac{e^{-\frac{\varepsilon}{kT}}}{2e^{-\frac{\varepsilon}{kT}}} = \frac{1}{2}$$

The electron with two equal energy states is equivalent to the two isomer model with two equi-energy isomers. The two electron states are equally probable; half of a set of electrons have the $+1/2$ quantum number.

Because the probabilities are formed as a ratio, the Boltzmann probabilities are often simplified by choosing a reference energy as the state of zero energy. For

M.E. Starzak, *Energy and Entropy*, DOI 10.1007/978-0-387-77823-5_11,
© Springer Science+Business Media, LLC 2010

the spin probabilities above, the exponentials in the numerator and denominator canceled because they were identical. Their energy could simply be selected as a reference energy of zero. In this case, the probability of the +1/2 state is

$$\frac{e^0}{e^0 + e^0} = \frac{1}{1+1} = 0.5$$

When the energies differ, one energy can be selected as the reference energy.

11.2 Unpaired Electrons in a Magnetic Field

If NO molecules with their unpaired electrons are placed in a magnetic field, the energy of the electrons depends on both the magnitude of the field and the spin quantum number. The common energy splits into two energies (Fig. 11.1).

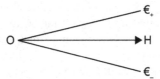

Fig. 11.1 Two electron energy states generated in a magnetic field

Electrons with spin -1/2 have an energy below the zero field reference energy $\varepsilon_- = -\varepsilon$, while the electrons with spin +1/2 increase their energy to $\varepsilon_+ = +\varepsilon$. The zero field energy is the reference. The Boltzmann factors for the negative and positive electron spins $e^{-\frac{\varepsilon_-}{kT}} = e^{-\frac{-\varepsilon}{kT}}$ are and

$$e^{-\frac{\varepsilon_+}{kT}} = e^{-\frac{+\varepsilon}{kT}}$$

respectively. The two-term partition function is

$$q = e^{\frac{\varepsilon}{kT}} + e^{-\frac{\varepsilon}{kT}}$$

The first Boltzmann factor is greater than 1. However, it is only a number; the probability is formed from a ratio of numbers and is always less than 1. The probability that the electron has the negative (lower) energy is

$$p_- = \frac{e^{+\frac{\varepsilon}{kT}}}{q} = \frac{e^{+\frac{\varepsilon}{kT}}}{e^{+\frac{\varepsilon}{kT}} + e^{-\frac{\varepsilon}{kT}}}$$

while the probability that the electron has the (positive) higher energy is

$$p_+ = \frac{e^{-\frac{\varepsilon}{kT}}}{q} = \frac{e^{-\frac{\varepsilon}{kT}}}{e^{+\frac{\varepsilon}{kT}} + e^{-\frac{\varepsilon}{kT}}}$$

The electron is in one of these two states with 100% probability:

$$\frac{e^{+\frac{\varepsilon}{kT}}}{q} + \frac{e^{-\frac{\varepsilon}{kT}}}{q} = \frac{q}{q} = 1$$

Numerical probabilities are determined for the special case where the magnetic field makes

$$|\varepsilon| = kT$$

The exponential arguments are then +1 and -1, respectively, and the probability of finding the lower energy state is

$$p_- = \frac{e^{+1}}{e^{+1} + e^{-1}} = \frac{2.73}{2.73 + .37} = 0.88$$

while the probability of the high-energy state is

$$p_+ = \frac{e^{-1}}{e^{+1} + e^{-1}} = \frac{.37}{2.73 + .37} = 0.12$$

The two probabilities sum to 1. $\varepsilon = kT$ strongly favors the population of the lower energy state. This kT provides insufficient energy to create a large population of the high-energy state.

The probabilities also give the total number of molecules that have a given spin in the magnetic field. For 0.88 probability, 0.88 mol/1 mol of electrons have the lower energy and 0.12 mol have the high energy. For n total moles

$$n_+ = np_+$$
$$n_- = np_-$$

respectively.

Since the probabilities are always ratios of Boltzmann factors, the choice of reference state is completely arbitrary. If the lower energy is selected as the reference energy (0), the upper energy is $+2\varepsilon$ higher. With $\varepsilon = kT$, the probability for the lower energy states are

$$p_- = \frac{e^0}{e^0 + e^{-2}} = \frac{1}{1 + .135} = 0.88$$

$$p_+ = \frac{e^{-2}}{e^0 + e^{-2}} = 0.12$$

These probabilities are temperature dependent. An increase in temperature moves the numerical values of the Boltzmann factors and the probabilities toward equality. At very high temperatures, both exponentials approach 1 and the probabilities become equal. There is sufficient energy to create either spin state.

Boltzmann probability always decreases with increasing energy. This equilibrium condition is violated by some non-equilibrium systems. A two-energy level laser system might undergo an inversion where the upper energy state population exceeds that of the ground state. Since the energies are fixed, the inversion has been explained with a negative absolute temperature that reverses the arguments in the Boltzmann factors. Negative temperature is simply a construct. The Boltzmann factor is fully valid only for systems at equilibrium. The laser is a driven system.

A system with three possible energies

$$\varepsilon, 0, -\varepsilon$$

has a partition function

$$q = e^{-\frac{\varepsilon}{kT}} + e^{\frac{0}{kT}} + e^{-\frac{-\varepsilon}{kT}}$$

The probabilities for finding the negative, 0, and positive energies are

$$p_- = \frac{e^{+\frac{\varepsilon}{kT}}}{q}$$

$$\#p_0 = \frac{1}{q}$$

$$p_+ = \frac{e^{-\frac{\varepsilon}{kT}}}{q}$$

respectively. For $|\varepsilon| = kT$, the probabilities for the three states are

$$p_- = \frac{2.73}{4.10} = 0.66$$

$$p_0 = \frac{1}{4.10} = 0.24$$

$$p_+ = \frac{0.37}{4.10} = 0.09$$

The ratios of probabilities for adjacent energy states are 0.37 reflecting the common energy difference between them.

11.3 The Average Energy

A two-state model where each state has the same energy ε' has an average energy ε' since this is the only energy possible. In a magnetic field, when $|\varepsilon| = kT$, 88% of the molecules have the lower energy and the average energy is closer to − but not equal to it since the remaining 0.12 high energy electrons raise the average.

The energy averaging procedure is like averaging examination grades. In a class of 100 students, 25 students earn a grade of 80, 50 earn a grade of 90, and the final 25 earn a grade of 100. The examination average is the sum of all grades divided by the number of students. For 25 80's, 50 90's, and 25 100's

$$G = \frac{25 \times 80 + 50 \times 90 + 25 \times 100}{25 + 50 + 25} = \frac{9000}{100} = 90$$

This average is a probability equation. The probability that one student received 80 is 0.25 (25/100). The probability a single student received 90 is 0.5 (50/100). The 25 students who received 100 generate a probability of 0.25 (25/100). The average grade is

$$<G> = \frac{25}{100}80 + \frac{50}{100}90 + \frac{25}{100}100 = p_{80}80 + p_{90}90 + p_{100}100$$
$$= 1/4\,(80) + 1/2\,(90) + 1/4\,(100) = 90$$

In a class of five students who receive grades of 70, 80(2 students), 90, and 100, the probabilities are

$$p_{70} = \frac{1}{5} \quad p_{80} = \frac{2}{5} \quad p_{90} = \frac{1}{5} \quad p_{100} = \frac{1}{5}$$

and the class average is

$$<G> = \sum p_i G_i = 0.2\,(70) + 0.4\,(80) + 0.2\,(90) + 0.2\,(100) = 84$$

The average need not correspond to one of the actual grades.

The probabilities convert the statistical behavior of the 100 students to the behavior of a single "average" student. Each student has a 25% chance of receiving an 80, a 50% chance of receiving a 90, and a 25% chance of receiving 100. These statistics improve with increasing numbers of matched students.

The set of probabilities can be used to determine any other functions of grades. For example, the average squared grade is

$$<G^2> = p_{80}\left(80^2\right) + p_{90}\left(90^2\right) + p_{100}\left(100^2\right)$$

The average energy for the two-state system with

$$|\varepsilon| = kT$$

and probabilities

$$p_- = 0.88 \qquad p_+ = 0.12$$

at 300 K, the energies for the two states are

$$\varepsilon_- = -kT = -1.38 \times 10^{-23}\,(300) = -4 \times 10^{-21}\,\text{J molecule}^{-1}$$
$$\varepsilon_+ = +kT = +1.38 \times 10^{-23}\,(300) = +4 \times 10^{-21}\,\text{J molecule}^{-1}$$

respectively. The average energy is the sum of the products of the probability that a molecule has a certain energy and its energy

$$<\varepsilon> = 0.88 \left(-4 \times 10^{-21}\right) + 0.12 \left(+4 \times 10^{-21}\right) = (-3.52 + 0.48) \times 10^{-21}$$
$$= -3.04 \times 10^{-21} \, J \, molecule^{-1}$$

or

$$<\varepsilon> = (-kT)\,(0.88) + (+kT)\,(0.12) = -0.76kT$$

The average value is negative and relatively close to the energy of the negative spin state because the larger percentage of the electrons have this energy. The 12% of molecules with positive spin raise the average 0.48×10^{-21} J above the energy of the lower state.

$<\varepsilon>$ is the average energy for a single molecule. Multiplication by N_a gives the average energy per mole

$$<E> = N_a \langle \varepsilon \rangle$$

For this example, the average energy is

$$<E> = \left(6.02 \times 10^{23}\right) \left(-3.04 \times 10^{-21}\right) = -1830 \, J \, mol^{-1}$$

11.4 A Differential Expression for Average Energy

For a two-state system with energies ε_- and ε_+, the average energy expression is

$$<\varepsilon> = \frac{e^{-\frac{\varepsilon_-}{kT}} \varepsilon_- + e^{-\frac{\varepsilon_+}{kT}} \varepsilon_+}{e^{-\frac{\varepsilon_-}{kT}} + e^{-\frac{\varepsilon_+}{kT}}}$$

The numerator is similar to the partition function in the denominator. Each term in the numerator is multiplied by a different energy. This numerator is generated directly by differentiating the denominator with respect to

$$\beta = \frac{1}{k_b T}$$

The partition function

$$q = e^{-\beta \varepsilon_-} + e^{-\beta \varepsilon_+}$$

is differentiated with

$$-\frac{\partial}{\partial \beta}$$

to give terms

$$e^{-\beta \varepsilon_i} \, [+\varepsilon_i]$$

and the negative derivative of the partition function

$$\frac{\partial q}{\partial(-\beta)} = -\frac{\partial q}{\partial\beta} = -\frac{\partial}{\partial\beta}\left[e^{-\beta\varepsilon_-} + e^{-\beta\varepsilon_+}\right] = -e^{-\beta\varepsilon_-}(-\varepsilon_-) + e^{-\beta\varepsilon_+}(-\varepsilon_+)$$

$$= \varepsilon_- e^{-\beta\varepsilon_-} + \varepsilon_+ e^{-\beta\varepsilon_+}$$

is the numerator in the average energy expression.

The average energy is

$$<\varepsilon> = \frac{-\frac{\partial q}{\partial\beta}}{q}$$

or

$$<\varepsilon> = -\frac{\frac{\partial q}{q}}{\partial\beta} = -\frac{\partial\ln(q)}{\partial\beta}$$

This differential expression is applicable to any partition function for discrete or continuum energies. If the partition function can be added or integrated to give a compact function, that function is differentiated to give the average energy. This is particularly important when the partition functions involve integrals over a continuum of energies.

Some partition functions for different types of independent energies are products of partition functions. Independent rotational and vibrational partition functions are products

$$q_{rot}q_{vib}$$

The logarithm is a sum

$$\ln(q_{rot}q_{vib}) = \ln(q_{rot}) + \ln(q_{vib})$$

Each term, when differentiated, gives an average energy

$$<\varepsilon> = <\varepsilon_{rot}> + <\varepsilon_{vib}>$$

11.5 Average Entropies and Free Energies

The canonical ensemble is a mixture of molecules with different energies. If molecules with different energy are considered distinguishable isomers, the system has a distribution of isomers. The two energy model is a mixture of two such energy isomers. If the isomers are distinguishable, the system has an entropy of mixing determined from the probabilities for each species.

The "mixture" of molecules with $\varepsilon_- = -\varepsilon$ and $\varepsilon_+ = \varepsilon$, i.e., two isomers, has a single particle entropy of mixing

$$S = -k\sum X_i \ln X_i = -R[X_i \ln X_1 + X_2 \ln X_2]$$

$$= k_b[p_1 \ln(p_1) + p_2 \ln(p_2)]$$

The mole fractions are equivalent to the probability that a given "isomer" is chosen when a single molecule is selected.

For two energy isomers, the entropy is

$$S = -k_b \sum_{i=1}^{2} X_i \ln X_i = -R[X_1 \ln X_1 + X_2 \ln X_2]$$

For

$$\varepsilon_- = -kT \quad \varepsilon_+ = +kT$$

And probabilities $p_- = 0.88$ and $p_+ = 0.12$, the entropy for this system

$$S = -k\left[\sum_{i=1}^{N} p_i \ln p_i\right] = -k[0.99 \ln 0.99 + 0.12 \ln 0.12] = 0.366k$$

is less than the equal probability distribution

$$S = k_b \ln(2) = 0.69k_b$$

The equal probability system has the largest entropy.

The Boltzmann probabilities

$$p_i = \frac{e^{-\frac{\varepsilon_i}{kT}}}{q}$$

are substituted into the entropy of mixing for the two-state model entropy

$$<s> = -k_b[p_+ \ln p_+ + p_- \ln p_-] = -k_b\left[\frac{e^{-\beta\varepsilon_+}}{q} \ln\left[\frac{e^{-\beta\varepsilon_+}}{q}\right] + \frac{e^{=\beta\varepsilon_-}}{q} \ln\left[\frac{e^{-\beta\varepsilon_-}}{q}\right]\right]$$

and separate as

$$\ln p_i = \ln\left[e^{-\frac{\varepsilon_i}{kT}}\right] - \ln q = -\frac{\varepsilon_i}{kT} - \ln q$$

The entropy

$$\frac{<S>}{k_b} = -[p_+ \ln p_+ + p_- \ln p_-] = -p_+\left[-\frac{\varepsilon_+}{kT}\right] + p_+\ln q - p_-\left[-\frac{\varepsilon_-}{kT}\right] + p_- \ln q$$

separates into groups of terms in ε and in $\ln(q)$, respectively. Since the probabilities sum to unity

$$p_+ \ln q + p_- \ln q = [p_+ + p_-] \ln(q) = \ln q$$

The remaining terms give the average molecular energy

$$<\varepsilon> = p_+\varepsilon_+ p_-\varepsilon_-$$

and the entropy is

$$\frac{<s>}{k} = \frac{<\varepsilon>}{kT} + \ln q$$

Both sides are now multiplied by k to give an average entropy

$$<s> = +\frac{<\varepsilon>}{T} + k\ln q$$

The molar Helmholtz free energy

$$A = E - TS$$

is converted to a molecular free energy and solved for the molecular entropy

$$a = \varepsilon - Ts$$
$$s = \varepsilon/T - a/T = <\varepsilon>/T - <a>/T$$

Comparing this with the distribution entropy

$$<s> = <\varepsilon>/T + k\ln(q)$$

gives

$$<a> = -kT\ln(q)$$

The energy and entropy equations can be used to show

$$\beta = 1/kT$$

The change in entropy with energy

$$dE = TdS - PdV = TdS$$
$$dS = dE/T$$
$$dS/dE = 1/T$$

is compared with the ratio

$$\frac{d<S>}{d<\varepsilon>} = \frac{1}{T}$$

Since

$$<s> = k\beta<\varepsilon> + k\ln(q)$$
$$1/T = d<S>/d<\varepsilon> = d[k\beta<\varepsilon> + k\ln(q)]/d<\varepsilon> = k + 0$$

and $\beta = 1/kT$.

11.6 The Chemical Potential

The chemical potential is an energy associated with the addition or subtraction of particles from a system. The chemical potential is a free energy per mole or a free energy per particle. The energy change in the system is the product

$$\mu dN$$

and the thermodynamic internal energy change is

$$dE = TdS - PdV + \mu dN$$

Since

$$dE = \left(\frac{\partial E}{\partial S}\right)_{V,N} dS - \left(\frac{\partial E}{\partial V}\right)_{S,N} dV + \left(\frac{\partial E}{\partial N}\right)_{S,V} dN$$

the thermodynamic chemical potential is

$$\mu = \left(\frac{\partial E}{\partial N}\right)_{S,V}$$

at constant volume and entropy.

A system contains two molecules (A_1 and A_2) that absorb discrete quanta of energy, ϵ. The entire system has two such quanta so only three states are possible: $(A_1, A_2) = (2, 0)$, $(0, 2)$, and $(1, 1)$. The quanta are indistinguishable (a Bose–Einstein system).

The entropy of this system is

$$S = k\ln(\Omega) = k\ln(3)$$

The chemical potential is the energy change on adding a particle without changing V or S. If a third molecule A_3 with no energy (0 quanta) is added to the system at constant volume, the total energy of the system is unchanged ($2\{\epsilon\}$). However, the number of states increases to 6; $(A_1, A_2, A_3) = (2, 0, 0)$, $(0, 2, 0)$, $(0, 0, 2)$, $(1, 1, 0)$, $(1, 0, 1)$, and $(0, 1, 1)$. The entropy increases to

$$S = k_B \ln(\Omega) = k_B\ln(6)$$

The definition of the chemical potential requires constant entropy. The addition of A_3 has not changed the internal energy but has changed the entropy.

The proper conditions are met using a hypothetical situation. A_3 is added to the system with a negative quantum, i.e., an energy, $-\varepsilon$. When the molecule is added the total quanta in the system decrease from 2 to 1. The internal energy of the system decreases. At the same time, only one quantum defines the states of the system.

Only the following three states are now possible: $(A_1, A_2, A_3) = (1, 0, 0)$, $(0, 1, 0)$, and $(0, 0, 1)$. The entropy is constant on addition of the negative energy molecule

$$S = k_b \ln(\Omega) = k_b \ln(3)$$

The conditions to define a chemical potential on the addition of one particle are now satisfied. The energy has decreased by

$$-\varepsilon$$

at constant volume and entropy. The chemical potential is the change in energy divided by the change in molecules at constant entropy and volume

$$\mu = \left(\frac{\partial E}{\partial N}\right)_{S,V} = \frac{-\varepsilon}{1} = -\varepsilon$$

The chemical potential, an intensive quantity, decreases on the addition of a molecule. If this energy is inserted into a Boltzmann factor

$$\exp\left[-(-\beta\varepsilon)\right] = \exp(\beta\mu)$$

The Boltzmann factor in chemical potential has a positive argument.

The Boltzmann factor increases with particles. Two particles in the system have a factor

$$\left[\exp(\beta\mu)\right]^2$$

The Boltzmann factor with positive argument is consistent with the thermodynamic expression for free energy (for one particle)

$$\mu = kT \ln(c)$$

Exponentiating gives

$$\exp(+\beta\mu)$$

The probability for N particles is proportional to

$$e^{+\frac{\mu N}{k_B T}}$$

11.7 Multi-particle Systems

Two independent identical molecules each have their own partition function. A new partition function, Q, generated from all total energies of the system is equivalent to the product of the two single molecule partition functions.

For two molecules with two-energy levels, ε_1 and ε_2, each molecule has a partition function,

$$q = e^{-\beta\varepsilon_1} + e^{-\beta\varepsilon_2}$$

The total energies for the two molecules are

(1) molecules 1 and 2 with energies ε_1;
(2) molecules 1 and 2 with energies ε_2;
(3) molecule 1 with energy ε_1 and molecule 2 with energy ε_2; and
(4) molecule 1 with energy ε_2 and molecule 2 with energy ε_1.

The last two states have the same energy

$$\varepsilon_1 + \varepsilon_2 = E_{12} = \varepsilon_2 + \varepsilon_1$$

so their Boltzmann factors are identical and can be combined. The two states with identical energy are degenerate states. The combined energy partition function

$$Q = e^{-\beta E_{11}} + 2e^{-\beta E_{12}} + e^{-\beta E_{22}}$$

has $E_{11} = 2\varepsilon_1$, $E_{12} = \varepsilon_1 + \varepsilon_2$, and $E_{22} = 2\varepsilon_2$.

This result can be obtained by squaring the single particle partition function

$$Q^2 = [\exp(-\beta\varepsilon_1) + \exp(\beta\varepsilon_2)]^2$$
$$\exp(-2\beta\varepsilon_1) + 2\exp(-\beta\{\varepsilon_1 + \varepsilon_2\}) + \exp(-2\beta\varepsilon_2)$$

The Boltzmann probabilities include both molecules. For both molecules with ε_1

$$p_{11} = \frac{e^{-\beta E_{11}}}{e^{-\beta E_{11}} + 2e^{-\beta E_{12}} + e^{-\beta E_{22}}}$$

The probability of molecules with ε_1 and ε_2 in either order requires the degeneracy

$$p_{12} = \frac{2e^{-\beta E_{12}}}{Q}$$

The two molecule partition function is the product of the two single particle partition functions

$$Q = q^2$$

For N independent molecules, the system partition function is

$$Q = q^N$$

The energy for the N molecules is

$$E = -\left(\frac{\partial \ln Q}{\partial \beta}\right) = \left(\frac{\partial \ln q^N}{\partial \beta}\right) = -N\left(\frac{\partial \ln q}{\partial \beta}\right) = N<\varepsilon>$$

Each molecule contributes its average energy to the total.

The partition function is a sum of numbers, i.e., a weighted count of states. If the particles are distinguishable, their order is incorporated into this sum. The two molecule system had one doubly degenerate state when one molecule had energy ε_1 and the second had energy ε_2. If the molecules are indistinguishable, this would be considered one state. The partition function is converted from distinguishable to indistinguishable by dividing by the number of orderings possible (2!)

$$Q_{ind} = q^2/2!$$

N indistinguishable particles with $N!$ Orderings have a corrected (for indistinguishabilty) Boltzmann partition function

$$Q_{ind} = q^N/N!$$

$N!$ does not change the average energy

$$\ln Q = \ln q^N - \ln N! = N \ln q - \ln N!$$
$$<E> = -N\left(\frac{\partial l \ln q}{\partial \beta}\right)$$

since $N!$ is constant.

11.8 Energy Manifolds

A single molecule can have translational, rotational, vibrational, and electronic energies. To a good first approximation, these energy manifolds are independent. The partition function for the rotational energies and the partition function for vibrational energies are generated separately. The product of all these independent energy partition functions within the single molecule are multiplied to produce the full partition function for a single molecule

$$Q(1 \text{ molecule}) = q_{trans}q_{rot}q_{vib}q_{elec}$$

A molecule with only two possible rotational energies, ε_{r1} and ε_{r2} and two possible vibrational energies ε_{r2} and ε_{v2} gives rotational and vibrational partition functions

$$[e^{-\beta\varepsilon_{r1}} + e^{-\beta\varepsilon_{r2}}]$$
$$[e^{-\beta\varepsilon_{v1}} + e^{-\beta\varepsilon_{v2}}]$$

Energies are added to get the total molecular energy. One molecule now has four possible energy states and the partition function for the two molecule system is

$$Q = e^{-\beta\varepsilon_{r1}}e^{-\beta\varepsilon_{v1}} + e^{-\beta\varepsilon_{r1}}e^{-\beta\varepsilon_{v2}} + e^{-\beta\varepsilon_{r2}}e^{-\beta\varepsilon_{v1}} + e^{-\beta\varepsilon_{r2}}e^{-\beta\varepsilon_{v2}}$$

Which factors as

$$Q(1 \text{ molecule}) = q_{\text{rot}}q_{\text{vib}}$$
$$[e^{-\beta\varepsilon_{r1}} + e^{-\beta\varepsilon_{r2}}][e^{-\beta\varepsilon_{v1}} + e^{-\beta\varepsilon_{v2}}]$$
$$= \exp[-\beta(\varepsilon_{r1} + \varepsilon_{v1})] + \exp[-\beta(\varepsilon_{r1} + \varepsilon_{v2})]$$
$$+ \exp[-\beta(\varepsilon_{r2} + \varepsilon_{v1})] + \exp[-\beta(\varepsilon_{r2} + \varepsilon_{v2})]$$

All possible energy combinations are generated by the product.

The product partition function generates a sum of independent translational, rotational, vibrational, and electronic energies. Real molecular systems can include some coupling between energy manifolds.

Problems

11.1 A quantum harmonic oscillator has large quanta so that only three-energy levels are populated. The molecule may have 0, 1, or 2 quanta exclusively. The Boltzmann factors for 0, 1, and 2 quanta are

$$e^{-\frac{\varepsilon_0}{kT}} = 1 \qquad e^{-\frac{\varepsilon_1}{kT}} = 0.2 \qquad e^{-\frac{\varepsilon_2}{kT}} = 0.04$$

a. Determine the partition function for this harmonic oscillator.
b. Determine the total probability that a molecule has either one or two quanta.
c. Determine the average number of quanta in a single harmonic oscillator molecule.

11.2 A molecule has only three allowed energy states, $\varepsilon_0 = 0$, $\varepsilon_1 = 1kT$, $\varepsilon_2 = 2kT$ with Boltzmann factors

$$e^{-\beta\varepsilon_0} = 1 \qquad e^{-\beta\varepsilon_1} = 0.4 \qquad e^{-\beta\varepsilon_2} = 0.2$$

a. For a system with 80 molecules, determine the numbers of molecule with energies $\varepsilon_0, \varepsilon_1, \varepsilon_2$ for the most probable distribution.
b. Determine the total average energy for one molecule using the most probable distribution from (a) in terms of kT.
c. Write an expression for the total number of states associated with the most probable distribution.

11.3 A molecule can exist as three isomers. The A isomer has a reference energy, $\varepsilon=0$ ($\exp(0/kT) = 1$. The B isomer has an energy, $\varepsilon_1 = 0.5kT[\exp(-0.5kT/kT) = 0.6]$ and the C isomer has an energy, $\varepsilon_2 = 0.9kT[\exp(-0.9kT/kT) = 0.4]$ at a temperature T.

a. Determine the probability of selecting an A isomer.
b. Determine the average energy for a single molecule.

c. Determine the entropy and Helmholtz free energy for the single molecule if the temperature is 300 K and $k = 1.4 \times 10^{-23}$ J K^{-1} molecule^{-1}.

11.4 A system has energies
$\varepsilon_0, \varepsilon_1, \varepsilon_2$ Give

a. the probability of finding a molecule in the lower energy state;
b. the average energy of the system;
c. the Helmholtz free energy for one molecule; and
d. the partition function Q for 10 indistinguishable particles.

Chapter 12
Multi-state Systems

12.1 The Harmonic Oscillator

The number of Boltzmann factors in the partition function increases with increasing energy states. The higher energy states might have small Boltzmann factors but are included for completeness. This leads to infinite sums that can sometimes be reduced to a simple partition function to calculate average energy, entropy, and free energy.

A classical vibrating molecule obeys Hooke's law for a displacement x–0

$$F = -kx$$

All displacements x are possible to give a continuum of possible energies.

A quantum mechanical oscillator, by contrast, accepts only discrete quantized energies that obey the equation

$$\varepsilon_v = \left(v + \frac{1}{2}\right) h v = \left(v + \frac{1}{2}\right) \frac{h}{2\pi} \sqrt{\frac{k}{\mu}} = (v+1/2)\, hv$$

$$v = 0,1,2,\ldots$$

For vibration of two masses m_1 and m_2 connected by a bond, μ is a reduced mass

$$1/\mu = 1/m_1 + 1/m_2$$

The mathematics is equivalent to a single mass μ bound to a fixed wall. k is the Hooke's law force constant. h is Planck's constant and v is the vibrational frequency that determines the energy.

The energy of the vth energy level is

$$\varepsilon_v = \varepsilon_{\frac{1}{2}} + vh\,v$$

Each term has energy $\varepsilon_{1/2} = \frac{1}{2}hv$ and the energy for each level can be broken into a constant term plus a v-dependent energy. Since the constant energy appears in each term, it can be factored to give a variable energy set

$$\varepsilon_v = hv(v + 1/2)$$

M.E. Starzak, *Energy and Entropy*, DOI 10.1007/978-0-387-77823-5_12,
© Springer Science+Business Media, LLC 2010

The constant $h\nu/2$ is then added to the average energy determined for the $\mathcal{E}_v$. The partition function for these energies is

$$q = \sum_{v=0}^{\infty} e^{-\frac{vh\nu}{k_bT}}$$

or

$$q = \sum_{v=0}^{\infty} x^v$$

with

$$x = \exp(-\beta h\nu)$$

Since

$$\frac{1}{1-x} = 1 + x + x^2 + \ldots = \sum_{i=0}^{\infty} x^i$$

The infinite sum reduces to

$$q = \frac{1}{1-x} = \frac{1}{1-e^{-\beta h\nu}}$$

Its logarithm

$$\ln q = \ln \frac{1}{1-e^{-\beta\varepsilon}} = -\ln\left(1 - e^{-\beta\varepsilon}\right)$$

determines the average energy

$$<\varepsilon> = -\frac{\partial \ln q}{\partial \beta}$$

$$= \frac{\partial}{\partial \beta} \ln\left[1 - \exp(-\beta h\nu)\right] = \left[1 - \exp(-\beta h\nu)\right]^{-1} \exp(-\beta h\nu)(h\nu)$$

The numerator and denominator are both multiplied by

$$e^{+\beta h\nu}$$

to give

$$<\varepsilon> = \frac{h\nu}{e^{\beta h\nu} - 1}$$

and a total energy

$$<\varepsilon> = \frac{h\nu}{2} + \frac{h\nu}{e^{\beta h\nu} - 1}$$

12.2 The Classical Limit

The average energy for a quantum harmonic oscillator for quantized energies, $h\nu$, differs from the predictions of classical equipartition theory

$$\varepsilon = 2\frac{1}{2}(kT) = kT$$

for one molecule. The difference is accentuated at low temperatures where the quantum partition function has few significant terms and produces very low energies. At higher temperatures, the quantum average energy approaches the classical value. The exponential argument

$$\varepsilon/kT$$

decreases with temperature. The exponential is approximated by the first two terms of a Taylor expansion:

$$e^{\beta\varepsilon} = 1 + \beta\varepsilon$$

With this approximation, the average energy expression reaches the classical limit

$$< \varepsilon >= \frac{h\nu}{1 + \frac{h\nu}{kT} - 1} = \frac{h\nu}{\frac{h\nu}{kT}} = kT$$

independent of the vibrational frequency.

This classical limit represents an upper bound for the harmonic oscillator energy when there is enough thermal energy (kT) to populate the discrete energy levels.

The heat capacity for a classical oscillator is the derivative of the average energy with respect to temperature:

$$C_v = \left(\frac{\partial < \varepsilon >}{\partial T}\right)_V = \left(\frac{\partial (kT)}{\partial T}\right)_V = k$$

The heat capacity for the quantum oscillator is

$$C_v = \left(\frac{\partial < \varepsilon >}{\partial T}\right)_v$$

The derivative requires three chain rule operations. The derivative of the parenthetical term

$$\frac{d}{dT}\left(e^{h\nu/(kT)} - 1\right)^{-1} = -1\left(e^{h\nu/(kT)} - 1\right)^{-2}$$

is followed by the derivative of the term in brackets (the exponential) and then the derivative of the argument of the exponential

$$\frac{d}{dT}\left[e^{h\nu/(kT)} - 1\right] = e^{h\nu/(kT)}\frac{h\nu}{-kT^2}$$

to give a heat capacity

$$C_v = \frac{\frac{hv}{kT^2} hv e^{hv/(kT)}}{\left(e^{hv/(kT)} - 1\right)^2}$$

12.3 The Helmholtz Free Energy and Entropy

The Helmholtz free energy for the harmonic oscillator

$$< a >= -kT \ln q = +kT \ln \left(1 - e^{-\beta \varepsilon}\right)$$

determines the work that the single molecule can provide. The energy ε is constant. Free energy differences appear only for temperature differences.

The free energies when $\varepsilon = kT$ at 600 K for 600 K and 300 K are

$$<a(600)> = k(600) \ln(1 - e^{-1}) = 275\,k$$
$$<a(300)> = k(300) \ln(1 - e^{-2}) = 44\,k$$

$275k - 44k = 231k$ of the free energy within one molecule can be released to the surroundings as the system temperature decreases.

The entropy for the harmonic oscillator is determined from the thermodynamic relation

$$<s> = \frac{<\varepsilon> - <a>}{T}$$

as

$$<s> = -k \ln\left[1 - e^{-\beta h v}\right] + \frac{\frac{hv}{T}}{e^{\beta hv} - 1}$$

12.4 Einstein's Crystal Heat Capacity

The law of Dulong and Petit states that all atomic crystals have a heat capacity of $3R$ (25 J K^{-1}mol$^-$). This remarkable observation means that a crystal with heavy atoms, e.g., gold, has exactly the same ability to absorb heat as a crystal of light atoms such as lithium. A single atom has a heat capacity of $3k$.

An atom in a crystal differs from an atom in a gas since it is trapped by its neighbors and cannot wander freely. The atom can store energy, however, by moving within the cavity formed by its neighbors. These oscillatory motions are effectively vibrations. The total equilibrium energy of one atom in the crystal with three vibrational directions is

$$< \varepsilon >= 3kT$$

The heat capacity is determined as the derivative of this energy with respect to the temperature

$$C = d < \varepsilon > /dT = 3k$$

This constant heat capacity is the law of Dulong and Petit for a single atom.

Classical physics was unable to explain the temperature dependence of atomic heat capacities. While $C = 3k$ at higher temperatures, it shows a steep decrease with decreasing temperature and approaches zero at T approaches 0 K (Fig. 12.1).

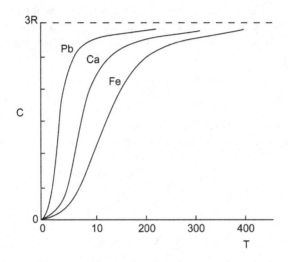

Fig. 12.1 Atomic crystal heat capacities as a function of temperature

Einstein invoked the quantum oscillator to explain the heat capacity decrease. The three atomic vibrations have a total quantum average energy

$$< \varepsilon >= \frac{3\varepsilon}{e^{\beta \varepsilon} - 1}$$

with $\varepsilon = h\nu$. The heat capacity of the crystal at any temperature is the temperature derivative of this average energy

$$C_v = 3k_B \left(\frac{h\nu}{k_B T} \right)^2 \frac{e^{\beta h\nu}}{\left(e^{\beta h\nu} - 1 \right)^2}$$

At the lower temperatures, fewer energy quanta are available to raise the oscillator energy by a discrete amount. The crystal absorbs less energy at these temperatures.

The Debye theory for heat capacities uses vibrations for the entire crystal rather than the vibrations of a single atom in a cavity to give a more accurate description of the decrease in heat capacity with temperature.

12.5 The Grand and Petit Canonical Partition Functions

The Boltzmann factor

$$e^{-\beta \varepsilon_{tot}}$$

generates an energy distribution. The grand and petit canonical partition functions generate both energy and particle distributions. Energies and chemical potentials are required to produce the distribution. The Boltzmann factors include both these energies

$$e^{-\beta \varepsilon} e^{\frac{\mu N}{kT}}$$

where N is the number of particles in the system.

The probability of finding a system with energy E and N particles is

$$p(E,N) = \frac{e^{-\beta E} e^{\beta \mu N}}{\Xi}$$

with

$$\Xi = \sum_E \sum_N e^{-\beta E} e^{\beta \mu N}$$

Although this double summation looks formidable, the most useful summations involve only a few terms. A petit canonical ensemble is used when a small number of ligands bind to a protein or surface site.

A protein has a partition function q_p incorporating all the energies of that protein. If these energies are not altered when ligand binds to the protein, the bare protein is assigned a reference energy of 0 and a Boltzmann factor $q_E = 1$. Since these protein energies and q_E are also present when a ligand binds, q_E cancels from any ratios, e.g., probabilities.

A binding ligand brings two energies to the protein as system. A chemical potential is necessary to bring the ligand from solution to the site. A binding energy, ε_b, is included for kinetic energy changes on docking. The petit partition function for a protein with one ligand binding site has only two terms, one for the protein and one for the protein with bound ligand

$$\xi = 1 + \exp\left(-\beta \varepsilon_b\right) \exp\left(+\beta \mu\right)$$

or

$$\xi = q_E + q_E e^{-\beta \varepsilon_b} e^{+1\beta \mu}$$

q_E cancels for probabilities. The probability that the protein has a bound ligand is

$$p_{ES} = \frac{q_E e^{-\beta \varepsilon_b} e^{\beta \mu}}{q_E \left[1 + e^{-\beta \varepsilon_b} e^{\beta \mu} \right]} = \frac{e^{-\beta \varepsilon_b} e^{\beta \mu}}{1 + e^{-\beta \varepsilon_b} e^{\beta \mu}}$$

or

$$p_{ES} = q_b \lambda / (1 + q_b \lambda)$$

where $q_p = \exp(\beta \varepsilon_b)$ $\lambda = \exp(+\beta \mu)$
The probability of ligand-free protein is

$$P_E = 1 / (1 + q_b \lambda)$$

while the probability for the protein–ligand complex is

$$p_{ES} = q_b \lambda / (1 + q_b \lambda)$$

For a protein with two identical binding sites, two bound ligands produce the Boltzmann factor

$$\lambda^2 = e^{-\beta 2\varepsilon_b} e^{\beta 2\mu}$$

and petit partition function

$$\xi = 1 + 2q_b \lambda + q_b^2 \lambda^2$$

The identical sites are degenerate. Their Boltzmann factors are identical for a single ligand binding to either site.

12.6 Multiple Sites

An ensemble of proteins with two binding sites has proteins with 0, 1, or 2 bound ligands. The average number of bound ligands lies between 0 where no ligands are bound to any protein to 2 where every protein has two ligands. The probabilities for a single protein are used to determine an average that lies between these two extremes.

The ligand-free protein is the reference Boltzmann state ($q_E = 1$). A ligand S binds to either identical site with binding energy, ε_b, and chemical potential, μ, to give two identical Boltzmann factors $q_{b\lambda}$. The state with two bound ligands has twice the energy and chemical potential to give a squared Boltzmann factor

$$q_b^2 \lambda^2$$

The petit canonical partition function

$$\xi = 1 + 2q_b\lambda + q_b^2\lambda^2 = (1 + q_b\lambda)^2$$

factors into two identical single site partition functions since each site is independent. Two sites with different energies would not reduce to a squared expression. A "system" that has N equal binding sites has partition function

$$\xi = (1 + q_b\lambda)^N$$

The partition function is the same for N independent sites on one protein or N single site proteins. The partition function recognizes only their independence. Expanding the partition function for three independent sites

$$(1 + q_b\lambda)^3 = 1 + 3q_b\lambda + 3(q_b\lambda)^2 + (q_b\lambda)^3$$

reveals the degeneracies. A protein with 2 ligands and 1 free site has degeneracy

$$3!/2!1! = 3$$

A protein with two different binding sites has energy and chemical potential ε_{b1} and μ_1 for site 1 and ε_{b2} and λ_2 for site 2 has four distinct terms

$$1 + q_{b1}\lambda_1 + q_{b2}\lambda_2 + q_{b1}\lambda_1 q_{b2}\lambda_2$$
$$= (1 + q_{b1}\lambda_1)(1 + q_{b2}\lambda_2)$$

Each independent site supplies its own partition function.

12.7 Binding Averages

The three site partition function

$$\xi = 1 + 3q_b\lambda + 3q_b^2\lambda^2 + q_b^3\lambda^3$$

has probabilities for 0,1,2,3 bound ligands

$$p_0 = 1/\xi$$
$$p_1 = 3q_b\lambda/\xi$$
$$p_2 = 3q_b{}^2\lambda^2/\xi$$
$$p_3 = q_b{}^3\lambda^3/\xi$$
$$\xi = 1 + 3q_b\lambda + 3q_b^2\lambda^2 + q_b^3\lambda^3$$

The average number of ligands bound to a single protein lies between 0 and 3 for this protein. The average substrate bound is the sum of the products of the probabilities times the number of substrate for that state

$$< n > \, = 0p_0 + 1p_1 + 2p_2 + 3p_3 = \frac{0(1) + 1\,(3q_b\lambda) + 2\left(3q_b^2\lambda^2\right) + 3\left(q_b^3\lambda^3\right)}{1 + 3q_b\lambda + 3q_b^2\lambda^2 + q_b^3\lambda^3}$$

$$= \frac{3q_b\lambda\left[1 + 2q_b\lambda + q_b^2\lambda^2\right]}{(1 + q_b\lambda)^3} = \frac{3q_b\lambda}{1 + q_b\lambda}$$

The averaging is the probability for binding at one independent site

$$\frac{q_b\lambda}{1 + q_b\lambda}$$

times *the total sites* (3).

$$\frac{3q_b\lambda}{1 + q_b\lambda}$$

For the three site enzyme, a Boltzmann factor

$$q_b\lambda = 1$$

gives

$$< n > \, = \frac{0 + (1)(3) + (2)(3) + (3)(1)}{1 + 3 + 3 + 1} = \frac{12}{8} = 1.5$$

Since a bound site is as likely as an empty site with this Boltzmann factor (they have the same energy), half of the available sites on the protein are bound.

12.8 A Differential Expression for Average Binding

The numerator for average number differs from the denominator only by the number of ligands in each term. This number is also a power of the Boltzmann factor. Differentiation of the partition function with λ reproduces the numerator when the result is multiplied by λ

$$\frac{\partial}{\partial\lambda}\left[1 + 3q_b\lambda + 3q_b^2\lambda^2 + q_b^3\lambda^3\right] = 0 + (1)3q_b + (2)3q_b\lambda^1 + 3q_b^3\lambda^2$$

$$\lambda\frac{\partial\xi}{\partial\lambda} = (1)q_b\lambda + (2)3q_b^2\lambda^2 + (3)q_b^3\lambda^3$$

This differential replaces the numerator for $<n>$

$$<n> = \frac{\lambda \frac{\partial \xi}{\partial \lambda}}{\xi} = \lambda \frac{\frac{\partial \xi}{\xi}}{\partial \lambda} = \lambda \frac{\partial \ln(\xi)}{\partial \lambda}$$

An enzyme with N sites has a partition function

$$\xi = (1 + \lambda q_b)^N$$

and

$$<n> = \lambda \frac{\partial \ln(1 + \lambda q_b)^N}{\partial \lambda} = \frac{N \lambda q_b}{1 + \lambda q_b}$$

A single site protein that binds different ligands A and B has a partition function

$$\xi = 1 + q_{bA} \lambda_A + q_{bB} \lambda_B$$

where each ligand has its own binding energy and chemical potential.

The average number of bound A and B are

$$<n_A> = \lambda_A d\ln(\xi)/d\lambda_A$$
$$<n_B> = \lambda_B d\ln(\xi)/d\lambda_B$$

12.9 Macroscopic Equilibria

A protein with a single binding site has probabilities

$$p_E = \frac{1}{1 + \lambda q} \qquad p_{ES} = \frac{\lambda q}{1 + \lambda q}$$

for E and ES. For total protein concentration c_t, the actual concentrations E and ES are

$$[E] = c_t \frac{1}{1 + \lambda q} \qquad [ES] = c_t \frac{\lambda q}{1 + \lambda q}$$

The concentrations determined using the equilibrium constant K and concentration c_t also have a ratio

$$K = \frac{[ES]}{[E][S]}$$
$$[ES] = K[E][S]$$
$$c_t = [E] + [ES]$$
$$= [E] + K[S][E]$$

the concentration of $[E]$ in terms of c_t

$$[E] = c_t \frac{1}{1 + K[S]}$$

also gives the concentration of $[ES]$

$$[EL] = c_t - [E] = c_t \frac{1 + K[S]}{1 + K[S]} - c_t \frac{1}{1 + K[S]} = c_t \frac{K[S]}{1 + K[S]}$$

The structure of these results parallels those for the microscopic statistical thermodynamic systems. Free protein is the reference state. The product $K[S]$ is the macroscopic equivalent of $q_b \lambda$. The Boltzmann factor in the chemical potential plays the role of a concentration at the microscopic level

$$C \alpha \lambda = \exp(+\beta \mu)$$

The Boltzmann factor for binding energy mimics the macroscopic equilibrium constant since strong binding to the site produces a large binding fraction consistent with a large equilibrium constant.

12.10 The Langmuir Adsorption Isotherm

A surface is an array of potential ligand binding sites for some ligand L. If binding at each site is independent of events at adjacent sites, surface binding analysis is done for a single site. The Langmuir model postulates one adsorbate molecule per surface site. Binding depends on binding energy and chemical potential. The Boltzmann factor when ligand binds

$$q_b \lambda = e^{-\beta \varepsilon_b} e^{\beta \mu}$$

plus reference as the bare site (1) gives

$$\xi = 1 + q_b \lambda$$

The grand canonical partition function for a surface with N binding sites is

$$Xi = \xi^N = (1 + q_b \lambda)^N$$

The average number of ligands at the site ($0 < <n> < 1$) is

$$<n> = f_S = \lambda \frac{\partial \ln(\xi)}{\partial \lambda} = \lambda \frac{\partial \ln(1 + q_b \lambda)}{\partial \lambda} = \frac{q_b \lambda}{1 + q_b \lambda}$$

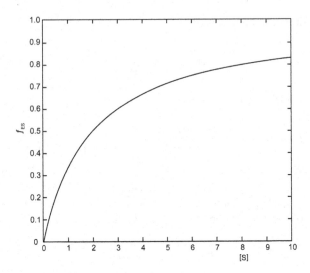

Fig. 12.2 Saturation in substrate binding

For N surface sites, the number with ligand is

$$< n_t >= \lambda \frac{\partial \ln (\Xi)}{\partial \lambda} = \lambda \frac{\partial \ln (1 + \lambda q)^N}{\partial \lambda} = \frac{N\lambda q}{1 + \lambda q}$$

$p_s = [S]/N$ is the probability for observing a substrate on a single site because each of the sites is independent. p_s approaches a limiting value of 100% as λ increases (Fig. 12.2). Since λq is the microscopic manifestation of $K[S]$ where $[S]$ is the adsorbate concentration in solution, this curve rises to 100% as the concentration of ligand in the solution above the surface increases.

12.11 The Brunauer–Emmett Teller (BET) Model

The Langmuir adsorption model limits binding to a single S on each binding site. The model can be extended to independent site, multiple substrate binding by adding a column of substrate to each of the binding sites. However, the first layer, the adsorbate, that binds directly to the surface, has different properties than subsequent layers that bind to the adsorbate molecules. For one site, the first adsorbate S is characterized by a specific binding energy and chemical potential to give the Boltzmann factor

$$\lambda q_b \equiv \lambda$$

Each subsequent substrate molecule sits on top of another substrate molecule with different binding energy and chemical potential. The Boltzmann factor for each of these S molecules is

$$\lambda' q' \equiv \lambda'$$

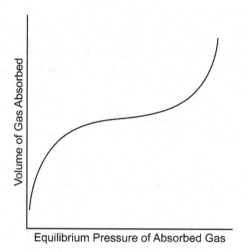

Fig. 12.3 The Brunauer–Emmett Teller (BET) model

The petit partition function includes the empty site and sites with 1, 2, 3, ...
substrates for an infinite series,

$$\xi = 1 + \lambda + \lambda\lambda' + \lambda\left(\lambda'\right)^2 + \ldots = 1 + \lambda \sum_{i=0}^{\infty} \left(\lambda'\right)^i = 1 + \frac{\lambda}{1 - \lambda'}$$

for $\lambda' < 1$.

The number bound per site can now be greater than 1. If S binds more strongly to
the surface than it does to other substrate, a plot of $<n>$ versus λ reaches a plateau,
then rises rapidly when λ' becomes large enough to induce multilayer binding
(Fig. 12.3).

When

$$\xi = 1 + \frac{\lambda q}{1 + \lambda q'}$$

the average number of bound molecules per site is evaluated as ($\lambda' = q_b\lambda$)

$$<N> = \lambda\frac{\partial \ln \xi}{\partial \lambda} = \lambda\frac{\partial \left[\ln\left(1 + \lambda q\left(1 - \lambda q'\right)^{-1}\right)\right]}{\partial \lambda}$$

$$= \frac{\frac{q\lambda}{1 - q'\lambda} + \frac{qq'\lambda^2}{(1 - q'\lambda)^2}}{1 + \frac{\lambda q}{1 - \lambda q'}} = \frac{q\lambda + \frac{qq'\lambda^2}{1 - q'\lambda}}{1 - \lambda q' + \lambda q} = \frac{q\lambda}{(1 - q'\lambda)(1 - q'\lambda - q\lambda)}$$

Problems

12.1 A large molecule with a reference energy 0 has two identical ion binding sites for ion A. One ion with chemical potential μ and binding energy ε binds to the protein.

a. Determine the partition function for the two site protein.
b. Determine the fraction of proteins that have no ions bound.

12.2 Diatomic molecular hydrogen lies on a surface where it can rotate. Assume that the interaction of this diatomic with the surface permits quantized non-degenerate energies $E_J = EJ$.

a. Write the petit canonical partition function for this system for a single molecule on the surface.
b. Determine the average energy for this system if $J = 1$ and 2 only.
c. Determine the Helmholtz free energy for 100 such molecules.

12.3 A very small metal surface has only three surface sites which are each capable of binding no more than 1 absorbate molecule A.

a. Label the three metal sites 1, 2, and 3 and show all possible states for the three site system.
b. Determine Ξ for this full system if there are no interactions between adsorbate molecules. The chemical potential for A is μ and is binding energy is E.

12.4 In one model of surface adsorption, only two "layers" are possible. The first molecule binds to the surface with an energy ε (Boltzmann factor $\lambda = e^{-\beta\varepsilon}$).A second molecule can then sit on top of the first adsorbed molecule with energy ε' (Boltzmann factor $\lambda' = e^{-\beta\varepsilon'}$).

a. Give the partition function for the single site with adsorbed molecules using the bare surface site as the reference state with zero energy.
b. Write an expression for the probability that the single site will have two adsorbed molecules.
c. Write an expression for the average number of adsorbed molecules at this site.
d. If the surface consists of three independent surface sites, write the partition function for the entire surface in terms of the partition function q for one site.

12.5 Two DNA chains of two nucleotides each, G_1–G_2 and C_1–C_2 interact at all sites.

a. Show all possible distinct configurations for the interactions between these two chains. Do not forget the "no interaction" configuration.

b. If each C–G interaction releases E_1, write the total partition function with degeneracies.

12.6 A molecule can exist as either an A or a B isomer. The A isomer is assigned a reference energy 0. Relative to this reference state, the B isomer has an energy ΔG per molecule.

a. Give an expression for the fraction of B isomers.
b. Only the B isomer has two sites that bind an S molecule when it is added to the solution. The A isomer has no such binding sites. If the additional Boltzmann factor for binding 1 S is

$$\lambda q_b = e^{\frac{\mu}{kT}} e^{-\frac{\varepsilon_b}{kT}}$$

write the full single molecule partition function which includes the A, B, BS, SB, and SBS forms of the isomer.
c. Determine the probability of finding B isomers in the solution. The B's include those with S bound.
d. Write an expression for the average number of S molecules bound to the isomer using the information from parts a, b, and c.

12.7 Evaluate the surface pressure for N molecules absorbed on a surface of M sites with N bound adsorbate molecules. For this specific absorption, use the partition function,

$$Q(N,M,T) = \{M!/N!(M-N)!\}\, q^N$$

Derive the equation

$$\mu/kT = -(\partial \ln Q/\partial N)_{M,T}$$

Develop an expression for the fraction of sites with bound molecules ($\theta = N/M$) from the partition function. This is the Langmuir absorption isotherm.

12.8 A protein can exist in two conformations, R and T, and the conformation T can bind ions I at two identical sites. Select R with reference energy $E = 0$. If T then has a relative energy E and the binding energy/ion is E_b,

a. Find the fraction of R molecules.
b. Determine the average number of bound ions.

The equilibrium constant for the R and T conformations

$$R \rightarrow T$$

is L. It is also the Boltzman factor for T.

12.9 A protein can exist in R or T conformations with an energy difference, $\Delta G = G_T - G_R$ The R conformation has two binding sites, site 1 and site 2, which are capable of binding ligands S_1 and S_2, respectively. S_1 cannot bind at site 2 and S_2 cannot bind at site 1. S_1 has a chemical potential μ_1 and binding energy E_1, while S_2 has a binding energy E_2 and a chemical potential μ_2.

a. List the possible states for the system and their energies.
b. Give the petit canonical partition function for this system.
c. Determine the fraction of proteins which would have the R conformation, i.e., they are not in the T conformation.
d. Develop an expression for the total bound substrate,$<S_1 +S_2>$.

12.10 A pair of electron exist in a triplet state with parallel spins or a single state with paired electrons. The triplet state splits into three separate energies in a magnetic field.

a. Determine the fraction of all pairs that are singlets.
b. If a magnetic field is turned on, the three equal triplet energies (0 energy reference) become $-\varepsilon,0,\ \varepsilon$. Determine the fraction of electron pairs with energy $+\varepsilon$

Chapter 13
Maxwell–Boltzmann Distributions

13.1 Distributions for Energy Continua

The Boltzmann factor

$$e^{-\beta\varepsilon_i}$$

describes a distribution for maximal states consistent with the available energy. If the separation between discrete energies is reduced to a continuum, the Boltzmann factor is proportional to the probability for an infinitesimal energy range. The partition function is an integration of Boltzmann factors over the range of continuous energies. The Boltzmann factor for $d\varepsilon$

$$\exp(-\beta\varepsilon)\,d\varepsilon$$

gives the partition function

$$q = \int \exp(-\beta\varepsilon)d\varepsilon$$

The discrete state equations for average energy and free energy are valid for the continuum partition function

$$<\varepsilon> = -\frac{\partial \ln q}{\partial \beta}$$

$$<a> = -kT\ln(q)$$

The probability that a classical oscillator has total *energy* ε. is proportional to the Boltzmann factor,

$$e^{-\frac{\varepsilon}{kT}} = e^{-\beta\varepsilon}$$

This two-dimensional energy (p,x) for the harmonic oscillator) has energy limits from zero to infinity, i.e., energy is always positive:

$$q = \int_0^\infty e^{-\beta\varepsilon}d\varepsilon$$

M.E. Starzak, *Energy and Entropy*, DOI 10.1007/978-0-387-77823-5_13,
© Springer Science+Business Media, LLC 2010

The differential energy $d\varepsilon$ gives the partition function units of energy. The numerator for average energy has units of energy squared for unit consistency.

The dimensionless probability for finding an energy from $\varepsilon \rightarrow \varepsilon + d\varepsilon$ is

$$p_{d\varepsilon} = \frac{e^{-\beta\varepsilon}d\varepsilon}{\int\limits_0^\infty e^{-\beta\varepsilon}d\varepsilon}$$

The probability of all possible energies is

$$\int\limits_0^\infty p(\varepsilon)\,d\varepsilon = \frac{\int\limits_0^\infty e^{-\beta\varepsilon}d\varepsilon}{\int\limits_0^\infty e^{-\beta\varepsilon}d\varepsilon} = 1$$

The partition function (not the energy) for the classical oscillator is

$$q = \int\limits_0^\infty e^{-\beta\varepsilon}d\varepsilon = \frac{-1}{\beta}e^{-\beta\varepsilon}\,\Big|_0^\infty = -\frac{1}{\beta}\left[e^{-\infty} - e^0\right] = \frac{1}{\beta} = kT$$

The probability for finding a molecule with an energy in the differential slice, $d\varepsilon$, is

$$p_{d\varepsilon} = p(\varepsilon)\,d\varepsilon = \frac{e^{-\beta\varepsilon}d\varepsilon}{\int\limits_0^\infty e^{-\beta\varepsilon}d\varepsilon} = e^{-\beta\varepsilon}\frac{d\varepsilon}{kT}$$

The ratio

$$d\varepsilon/kT$$

is a dimensionless variable of integration.

The probability of finding oscillators with energy from ε_1 to ε_2 is

$$P(\varepsilon_1 \rightarrow \varepsilon_2) = \int_{\varepsilon_1}^{\varepsilon_2} \exp(-\beta\varepsilon)\frac{d\varepsilon}{kT}$$

The average energy of a single classical harmonic oscillator is determined from the logarithm of the partition function in β format

$$\ln q = \ln(kT) = \ln\frac{1}{\beta} = -\ln\beta$$

$$<\varepsilon> = -\frac{\partial\ln q}{\partial\beta} = +\frac{\partial\ln\beta}{\partial\beta} = \frac{1}{\beta} = kT$$

as expected for the two degrees of freedom of the classical harmonic oscillator.

The free energy is

$$< a >= -kT \ln (kT)$$

13.2 Useful Integrals

The kinetic and potential energies of the classical harmonic oscillator

$$\varepsilon = \frac{mv^2}{2} + \frac{kx^2}{2}$$

each generate a Boltzmann factor. The kinetic energy factor is integrated over all possible velocities while the potential energy factor is integrated over all possible x. An ideal gas has no potential energy ($\exp(0) = 1$) and a three-dimensional integral gives the volume V of the container. A one-dimensional integral gives the length L of the line.

The one-dimensional translational kinetic energy

$$K.E. = \frac{mv^2}{2}$$

makes velocity, not energy, the integration variable. The Boltzmann factor has v^2 and range $-\infty \leq v_x \leq \infty$

$$q = \int_{-\infty}^{\infty} \exp\left(-\beta mv^2/2\right) dv_x$$

This integral appears for energies with a squared variable dependence, e.g., $kx^2/2$, $mv^2/2$. The integral for a general

$$e^{-ax^2}$$

with a constant is

$$\int_{-\infty}^{\infty} e^{-ax^2} dx = \sqrt{\frac{\pi}{a}}$$

The second major integral

$$\int_{0}^{\infty} e^{-ax^2} x dx = \frac{1}{2a}$$

is non-zero on the interval $0 \leq x \leq \infty$ for this odd product; it is zero on the interval $-\infty < x \leq \infty$.

The integral

$$\int_{-\infty}^{\infty} e^{-ax^2}\,dx = \sqrt{\frac{\pi}{a}}$$

is evaluated using the product of two identical integrals in different variables (x and y)

$$I^2 = \int_{-\infty}^{\infty} e^{-ax^2}\,dx \int_{-\infty}^{\infty} e^{-ay^2}\,dy$$

$$\int_{-\infty}^{\infty}\int_{-inf}^{\infty} e^{-a(x^2+y^2)}\,dxdy$$

The pair is simplified by a change to polar coordinates. The differential product $dxdy$ becomes

$$dxdy = rdrd\theta$$

and

$$x^2 + y^2 = r^2$$

r is integrated from 0 to ∞ and θ is integrated from 0 to $360° = 2\pi$ to span the full two-dimensional space.

The two integrals convert to

$$\int_0^\infty\int_0^{2\pi} e^{-ar^2} rdrd\theta = \int_0^\infty e^{-ar^2} rdr \int_0^{2\pi} d\theta$$

The integral over all angles is just 2π. The radial integral is integrable with the substitution

$$z = ar^2 \quad dz = 2ardr \qquad rdr = \frac{dz}{2a}$$

to give

$$\int_{z=0}^{z=\infty} e^{-z}\frac{dz}{2a} = \frac{1}{2a}[-e^{-z}]\Big|_0^\infty = +\frac{1}{2a}$$

$$I^2 = \frac{2\pi}{2a} = \frac{\pi}{a}$$

is the product of two identical integrals so

$$I = \sqrt{\frac{\pi}{a}}$$

Integrals such as

$$\int\limits_{-\infty}^{\infty} e^{-ax^2} x^2 dx$$

are evaluated by differentiating both sides of the integral

$$\int\limits_{-\infty}^{\infty} e^{-ax^2} dx$$

with respect to $-a$

For example,

$$\frac{-\partial}{\partial a}\left[\int\limits_{-inf}^{\infty} e^{-ax^2} dx\right] = -\frac{\partial}{\partial a}\sqrt{\pi}\,a^{-1/2}$$

$$\int\limits_{-\infty}^{\infty} x^2 e^{-ax^2}\Big] dx = -\sqrt{\pi}\left(-\frac{1}{2}\right)a^{-\frac{3}{2}} = (1/2a)\sqrt{\pi/a}$$

Other integrals with even powers of x, e.g., x^4, x^6, are solved by further differentiation of both sides of the integral.

For odd powers of x, e.g., x^3, x^5 with

$$\int\limits_{0}^{\infty} e^{-ax^2} x dx = \frac{1}{2a}$$

repeated differentiations produce the integrals in x^3, x^5, etc.

For example,

$$\int\limits_{0}^{\infty} e^{-ax^2} x^3 dx = \frac{1}{2a^2}$$

13.3 One-Dimensional Velocity Distribution

One-dimensional velocities range from $-\infty \leq v_x \leq \infty$ since the particles can move in either direction along x.

The squared velocity in the exponential produces a symmetrical decrease in probability for either positive or negative velocities. The resultant distribution is maximal at and centered about $v = 0$.

The most probable velocity occurs when the derivative with respect to v is 0

$$d\left(\exp\left(-\beta m v^2/2\right)\right) = 0$$
$$\exp\left(-\beta m v^2/2\right)(\beta m v) = 0$$
$$v = 0$$

The average squared energy is always positive and the integral is non-zero. The partition function with Boltzmann factor

$$\exp\left(\frac{-m v_x^2}{2}\right)$$

gives a probability

$$p\left(v_x\right) dv_x = \frac{e^{\left(-m v^2/2kT\right)} dv_x}{\int\limits_{-\infty}^{+\infty} e^{\left(-m v^2/2kT\right)} dv_x}$$

The partition function is the integral

$$\int\limits_{-\infty}^{\infty} \exp\left(-a x^2\right) = \sqrt{\frac{\pi}{a}}$$

with

$$a = m/2kT$$

i.e.,

$$q = \int\limits_{-\inf}^{\infty} e^{-a v^2} dv_x = \sqrt{\frac{\pi}{a}} = \sqrt{\frac{2\pi kT}{m}} = \sqrt{\frac{2\pi}{m\beta}}$$

The logarithm of partition function with β separates into two parts:

$$\ln q = \ln\left(\sqrt{\frac{2\pi kT}{m}}\right) = \ln\left(\frac{2\pi}{\beta m}\right)^{1/2} = \frac{1}{2}\ln\left(\frac{2\pi}{m}\right) - \frac{1}{2\ln(1/\beta)}$$

The average energy depends only on β

$$< \varepsilon > = -\frac{\partial\left(\frac{1}{2}\frac{\ln(\beta^{-1})}{\beta}\right)}{\partial\beta} = \frac{+1}{2}\frac{\partial\ln\beta}{\partial\beta} = \frac{1}{2\beta} = \frac{kT}{2}$$

The result is the equipartition of energy theorem for one-dimensional translation.

The average velocity is the product of an even ($\exp(-\beta mv^2)$) and odd (v_x) function. The product is odd to give 0 average velocity

$$< v_x >= \int_{-\infty}^{\infty} v_x p\,(v_x)\,dv_x = \frac{\int_{-\infty}^{\infty} e^{(-\beta mv_x^2/2)} v_x dv_x}{q} = \frac{\int_{-\infty}^{\infty} e^{(-\beta mv_x^2/2)} v_x dv_x}{\sqrt{\frac{2\pi k_b T}{m}}}$$

$$\int_{-\infty}^{\infty} e^{-av^2} v\,dv = \frac{-1}{2a} e^{-av^2} \Big|_{-\infty}^{\infty} = 0 - 0 = 0$$

The distribution widens as temperature increases (Fig. 13.1).

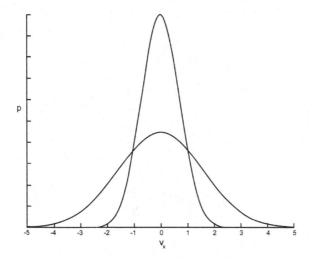

Fig. 13.1 The change of velocity distribution with increasing temperature

13.4 Mean Square Velocities

The one-dimensional velocity distribution has an average energy

$$< \varepsilon >= \frac{k_b T}{2}$$

while the average velocity is

$$< v_x >= 0$$

The zero average velocity does not give an average energy of zero since the energy is proportional to v^2 and this must be averaged

$$< \varepsilon > = < \frac{1}{2}mv_x^2 > = \frac{m}{2} < v_{x.}^2 >$$

This average square velocity is calculated from the average equipartition energy

$$< v_x^2 > = \frac{2 < \varepsilon >}{m} = \frac{2\frac{kT}{2}}{m} = \frac{kT}{m}$$

and square rooted for a root mean square (rms) velocity

$$v_{rms} = \sqrt{v_x^2} = \sqrt{\frac{kT}{m}} = \sqrt{\frac{RT}{M}}$$

$< v_x^2 >$ is also determined using the Boltzmann probabilities

$$< v^2 > = \int_{-\infty}^{\infty} p(v_x) \, v^2 dv_x$$

$$< v^2 > = \frac{\int_{-\infty}^{\infty} e^{(-mv^2/2kT)} v_x^2 dv_x}{\sqrt{\frac{2\pi kT}{m}}}$$

with

$$a = \frac{\beta m}{2} = \frac{m}{2kT}$$

$$< v^2 > = \frac{\frac{1}{2a}\sqrt{\frac{\pi}{a}}}{\sqrt{\frac{\pi}{a}}} = \frac{1}{2a} = \frac{2kT}{2m} = \frac{kT}{m}$$

as expected.

13.5 Two-Dimensional Distributions

A one-dimensional partition function has a single differential. A two-dimensional systems (velocities along the x and y axes) has a differential element

$$dv_x dv_y$$

Since the directions are independent when the potential energy $v = 0$, the two-dimensional system separates into two integrals with Boltzmann factors and differential elements in x and y, respectively

$$\int_{-\infty}^{\infty}\int_{-\infty}^{\infty} \exp\left(-\beta m v_x^2\right) \exp\left(-\beta m v_y^2/2\right) dv_x dv_y$$

$$= \int_{-\infty}^{\infty} \exp\left(-\beta m v_x^2/2\right) dv_x \int_{=\infty}^{\infty} \exp\left(-\beta m v_y^2/2\right) dv_y$$

$$= q^2 = \left(\sqrt{\frac{2\pi}{m\beta}}\right)^2 = \frac{2\pi}{m\beta}$$

This result is also obtained using a polar coordinate system with

$$v_r = \sqrt{\left(v_x^2 + v_y^2\right)}$$

and polar angle θ ($0 \le \theta \le 2\pi$) (dv_r is the infinitesimal radial length). The angular length depends on the radius as $v d\theta$ for an area element

$$v dv d\theta$$

The two Boltzmann arguments are combined to give a two-dimensional speed

$$V^2 = v_x^2 + v_y^2$$

The two-dimensional partition function is the product of $d\theta$ and du integrals

$$\int_0^{2\pi} d\theta \int_0^{\infty} \exp\left(-\beta m v^2/2\right) v dv = 2\pi \left[\frac{2}{2\beta m}\right] = \frac{2\pi}{\beta m}$$

equivalent to that obtained as the product of two one-dimensional partition functions.

The average energy is

$$< \varepsilon > = -\left(\frac{\partial\left[\ln\left(\frac{2\pi}{m}\right) - \ln\left(\beta\right)\right]}{\partial \beta}\right) = \frac{1}{\beta} = kT$$

13.6 Three-Dimensional Velocity Distributions

A three-dimensional velocity distribution uses the speed

$$v^2 = v_x^2 + v_y^2 + v_z^2$$

that defines a hollow sphere with radius v in this velocity three-dimensional space. The surface area of the sphere is

$$4\pi v^2$$

and the differential volume of the hollow sphere is this area times dv

$$4\pi v^2 dv$$

The partition function requires a single integral in the speed v. The factor of 4π is eliminated since it is common to all terms

$$q = \int_0^\infty e^{(-\beta mv^2/2)} v^2 du = \int_0^\infty e^{-av^2} v^2 dv = \frac{1}{2}\frac{1}{2a}\sqrt{\frac{\pi}{a}} = \frac{1}{4a}\sqrt{\frac{\pi}{a}}$$

$1/2$ is included for integration from 0 to ∞.
The average energy is

$$<\varepsilon> = -\frac{\partial\left[\ln(\pi)^{\frac{3}{2}} - \ln\beta^{\frac{3}{2}}\right]}{\partial\beta} = +\frac{3}{2}\frac{1}{\beta} = \frac{3}{2}kT$$

as expected for the three-dimensional system.

Since the three Cartesian velocity components are independent, this same partition function is also a product of three one-dimensional partition functions

$$q = q_{v_x} q_{v_y} q_{v_z} = \left[\sqrt{t}\frac{2\pi}{m\beta}\right]^3 = \left(\frac{2\pi k_b T}{m}\right)^{\frac{3}{2}}$$

The Boltzmann probability has partition function

$$q = (2\pi/\beta m)^{3/2}$$

to give

$$p(v)\,dv = \frac{e^{(-mv^2/2kT)} v^2 dv}{q}$$

Although the Boltzmann factor decreases as the speed increases, the volume of each concentric sphere as $v^2 dv$ increases with increasing v. These two opposing changes combine to give a maximum at an intermediate speed (Fig. 13.2).

The average energy for the three-dimensional system

$$\ln q = \ln\left(\frac{2\pi}{m}\right)^{3/2} - \frac{3}{2}\ln\beta$$

$$\varepsilon = -\frac{\partial\left(-\frac{3}{2}\ln\beta\right)}{\partial\beta} = \frac{3}{2\beta} = \frac{3kT}{2}$$

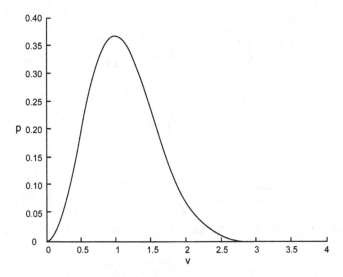

Fig. 13.2   3 dimensional velocity distribution

is again consistent with equipartition of energy.
   The speed probability distribution

$$p\left(v\right)dv = \frac{e^{\left(-mv^2/2kT\right)}v^2dv}{q}$$

is multiplied by the speed v and integrated to find $<v>$

$$<v>=\int_0^\infty p\left(v\right)vdv$$

The ratio

$$<v>=\frac{4\pi\int_0^\infty e^{-av^2}v^3dv}{4\pi\int_0^\infty e^{-av^2}v^2dv}$$

requires integrals

$$\int_0^\infty e^{-ax^2}x^3dx = -\frac{\partial\frac{1}{2a}}{\partial a} = \frac{1}{2a^2}$$

and

$$\int_0^\infty \exp\left(-ax^2\right)x^2dx = \frac{1}{4a}\sqrt{\frac{\pi}{a}}$$

with

$$a = \frac{m}{2k_b lT}$$

<v> is

$$< v >= \frac{4\pi\frac{1}{2a^2}}{q} = \frac{4\pi\frac{1}{2a^2}}{4\pi\frac{1}{2}\frac{1}{2a}\sqrt{\frac{\pi}{a}}} = \frac{2}{a}\sqrt{\frac{a}{\pi}} = \sqrt{\frac{4}{\pi a}} = \sqrt{\frac{8kT}{\pi m}}$$

An equivalent expression in the gas constant R and the molecular weight M (in kilograms) is

$$< v >= \sqrt{\frac{8k_b N_a T}{\pi m N_a}} = \sqrt{\frac{8RT}{\pi M}}$$

Each form gives units of meters per second.
The kinetic energy for each molecule depends on its average squared speed:

$$< v^2 >= \int_0^\infty p\left(v\right)v^2 du$$

$$< u^2 >= \frac{\int_0^\infty e^{-\frac{\mu^2}{2kT}}v^2v^2 du}{q}$$

The numerator is

$$\frac{\partial^2\int_0^\infty e^{-av^2}du}{\partial a^2} = \frac{\partial^2\frac{1}{2}\sqrt{\pi}a^{\frac{-1}{2}}}{\partial a^2}$$

$$\frac{1}{2}\frac{1}{2}\frac{3}{2}a^{\frac{-5}{2}} = \frac{3}{8}a^{-\frac{5}{2}}$$

to give

$$< v^2 >= \frac{\frac{3}{8}a^{\frac{-5}{2}}}{\frac{1}{4}a^{-\frac{3}{2}}} = \frac{3}{2a} = \frac{3k_b T}{m}$$

or

$$< v^2 >= \frac{3RT}{M}$$

The average kinetic energy is

$$< \varepsilon > = \frac{m}{2} < v^2 > = \frac{m}{2} \frac{3kT}{m} = \frac{3}{2} k_b T$$

The root mean square speed is the square root of this speed:

$$v_{\mathrm{rms}} = \sqrt{\frac{3kT}{m}} = \sqrt{\frac{3RT}{M}}$$

The larger speeds are weighted more heavily so the root mean square contains a factor of 3 while the average speed contains a factor

$$\frac{8}{\pi} = 2.55$$

Because of the shape of the speed distribution, the average speed and the root mean square speed differ from the speed of maximum probability. The most probable speed is determined by equating the derivative in v to 0

$$\frac{\partial p\,(v)}{\partial v} = 0 \qquad \frac{\partial e^{-av^2} v^2}{\partial v} = e^{-av^2}[-2av]v^2 + e^{-av^2}[2v]$$

to give

$$0 = -2av^3 + 2v \qquad v^2 = \frac{1}{a} = \frac{2kT}{m}$$

The most probable speed is the square root of this expression:

$$v_{\mathrm{mp}} = \sqrt{\frac{2kT}{m}} = \sqrt{\frac{2RT}{M}}$$

This speed is smaller than either the average speed or the root mean speed since both of these average quantities emphasize the larger speeds of this skewed speed distribution.

13.7 The Classical Harmonic Oscillator

A classical harmonic oscillator has kinetic and potential energies

$$E = \frac{1}{2} m v^2 + \frac{1}{2} k' x^2$$

where k' is the force constant. The Boltzmann factor separates as the product of Boltzmann factors in v and x:

$$e^{-\beta\left[\frac{1}{2}mv^2 + \frac{1}{2}kx^2\right]} = e^{-\frac{\beta mv^2}{2}} e^{-\frac{\beta kx^2}{2}}$$

The partition function has integrals over v and x

$$q = \int\limits_{-\infty}^{\infty} e^{-\frac{\beta mv^2}{2}} dv \int\limits_{-\infty}^{\infty} e^{-\frac{\beta kx^2}{2}} dx$$

The first integral is identical to the one-dimensional velocity distribution integral:

$$q_v = \sqrt{\frac{2\pi}{\beta m}}$$

The position partition function with $a = \beta k/2$ is

$$q_x = \int\limits_{-\infty}^{\infty} e^{-ax^2} dx = \sqrt{\frac{\pi}{a}} = \sqrt{\frac{2\pi}{\beta k}}$$

The logarithm of the product separates into sums

$$\ln(q_v q_x) = \ln(q_v) + \ln(q_x) = \frac{1}{2}\ln\left(\frac{2}{m}\right) - \frac{1}{2}\ln(\beta) + \frac{1}{2}\frac{\ln}{(2k)} - \frac{1}{2}\ln(\beta)$$

$$= \ln(\text{constant}) - \ln(\beta)$$

to give an average energy

$$<E> = -\frac{\partial}{\partial\beta}[\ln(\text{constant}) - \ln(\beta)] = +\frac{1}{\beta} = kT$$

Each squared term, integrated over its independent variable, gives an average energy $kT/2$, the equipartition of energy.

13.8 The Quantum Rotator

A classical diatomic rotator rotates about only 2 of its three rotational axes. Each of these rotations has a kinetic energy

$$E = \frac{1}{2}I\omega^2$$

where I is the moment of inertia of the rotator. The partition function for each rotation, determined by an integration over all possible angular velocities,

$$q_r = \int\limits_{-\inf}^{\infty} e^{(-\beta I \omega^2 / 2)} d\omega = \sqrt{\frac{2\pi}{\beta I}}$$

gives an average energy $kT/2$ for each independent rotation.

A quantum rotator has discrete energy levels

$$E_J = \varepsilon J (J + 1) \quad J = 0,1,2,\ldots$$

where ε is constant and the Jth quantum level is degenerate; there are $2J+1$ equal energies with quantum number J

$$(2J + 1) e^{-\beta \varepsilon J (J+1)}$$

The sum over these discrete Boltzmann factors can be approximated by an integral using the index dJ in place of the integer J because the small quantized rotational energies are very closely spaced. The partition function

$$q_r = \int\limits_0^{\infty} (2J + 1) e^{-\beta \varepsilon J (J+1)} dJ$$

is evaluated using the substitutions

$$y = J (J + 1)$$
$$dy = (2J + 1) dJ$$

to give

$$q_r = \int\limits_0^{\infty} e^{-\beta \varepsilon y} dy = -\frac{e^{-\beta \varepsilon y}}{\beta \varepsilon} \Big|_0^{\infty} = 0 + \frac{1}{\beta \varepsilon}$$

Although the partition function is different than that for the classical rotator with two rotational modes, the average energy is the same:

$$< E > = -\frac{\partial}{\partial \beta} [-\ln (\varepsilon) - \ln (\beta)] = \frac{1}{\beta} = kT$$

The hydrogen diatomic molecule has quantum properties that require refinement of its partition function. Each hydrogen atom has a nuclear spin of +1/2 or -1/2. The spin components of the two nuclei give two different hydrogen molecules: *ortho* hydrogen has both spin components equal, e.g., +1/2 and +1/2, while *para* hydrogen has nuclei with opposite spin components, e.g., +1/2 and -1/2. *Para* hydrogen is described by a single molecular wavefunction, while *ortho* hydrogen has three wavefunctions for spin components (+1/2,+1/2), (1/2, -1/2) + (-1/2, +1/2), and (-1/2, -1/2). The three *ortho* hydrogen molecules have an even nuclear symmetry, i.e., the nuclei can be exchanged without changing the sign of the wavefunction. The *para* hydrogen molecule is odd or negative in an interchange of the two nuclei.

The Pauli exclusion principle states that a valid wavefunction must be odd or antisymmetric. The total wavefunction for the molecule is a product of electronic, vibrational, rotational, and nuclear wavefunctions. The nuclear wavefunctions are symmetric or antisymmetric and the rotational wavefunctions can also be odd or even. The wavefunctions with even quantum number J (0,2,4,...) are even.

The integral partition function over all J, q, is halved when the integration is over either even or odd J

$$q_o = (q/2) = q_e$$

Since each of the *ortho* forms is equally probable, the total partition function for an equilibrium mixture of *ortho* and *para* hydrogen molecules is

$$Q_t = 3q_o + q_p = 3q/2 + q/2 = 2q$$

The total partition function is twice as large as the "classical" q.

13.9 Phase Space

The Maxwell–Boltzmann distributions use velocity, not energy, as the independent integration variable. The kinetic energy

$$mv^2/2$$

in momentum mv is

$$\frac{(m^2v^2)}{2m} = \frac{p^2}{2m}$$

Momentum and position play a central role in quantum mechanics where the uncertainty principle

$$\Delta x \Delta p \approx h$$

establishes a level of observability. The uncertainty principle also establishes a minimum size for area in a space with one x coordinate and an orthogonal p_x coordinate. This phase space area is used as a minimum state size.

A two-dimensional phase space for a vibrating diatomic molecule on a one-dimensional axis has x and p_x axes for the independent variables that give

$$\varepsilon = p^2/2m + kx^2/2$$

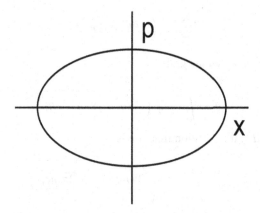

Fig. 13.3 A constant energy phase space trajectory for a harmonic oscillator

The velocity is 0 when the oscillator reaches its maximal positive and negative x (x_m) for one energy. $x=0$ when positive and negative p_x are maximal. The trajectory in phase space is an oval that passes through these four points (Fig. 13.3).

For a classical oscillator, an increase in energy produces a larger closed concentric trajectory. A quantum oscillator with discrete energies produces a set of discrete concentric trajectories in the phase space.

A three-dimensional system has three position coordinates (x, y, and z) and three momentum coordinates (p_x, p_y, and p_z). This six-dimensional space cannot be illustrated graphically but a fixed energy forces the oscillator to follow a closed trajectory in this space.

A complete one-dimensional partition function must involve both the p_x (or v_x) variable and the position (x) variable. Gas particles in one dimension have a potential energy zero and the spatial portion is integrated over the length of the line

$$\int_{-\infty}^{\infty} \exp\left(-\beta p^2 /2m\right) dp \int_0^L \exp\left(0\right) dx = \sqrt{\frac{2\pi m}{\beta}} L$$

The logarithm of this partition function would separate into an energy (p) and locational (x) component. This is the classical partition function for translation.

13.10 Quantized Phase Space

The quantum mechanical version of translation on a one-dimensional line of length L requires the discrete energies for a particle on a line

$$\varepsilon = h^2 n^2 /8mL^2 \quad n = 1,2,3,\ldots$$

Discrete energies are apparent when L is small. For larger L, the summed partition function

$$\sum_{n=1}^{\infty} \exp\left(-\beta h^2 n^2 / 8mL^2\right)$$

is well approximated by an integral

$$q = \int_0^{\infty} \exp\left(-\beta h^2 n^2 / 8mL^2\right) dn$$

with $a = \beta h^2 / 8mL^2$, the partition function is

$$q = \frac{1}{2}\sqrt{\frac{\pi}{a}} = \frac{1}{2}\sqrt{\frac{8\pi mL^2}{\beta h^2}} = \frac{\sqrt{2\pi m/\beta}}{h}L$$

This quantum energy partition function is almost identical to the partition function for a classical translating particle confined to length L with zero potential energy. The numerator is the volume in the classical phase space. The equation divides this volume by the uncertainty volume h in phase space

$$\Delta x \Delta p \approx h$$

This smallest "observable" volume gives the number of states for the system. A classical partition function can be converted into a number of possible states by dividing each two-dimensional (p,X) partition function by h.

If the trajectories of a large number of different oscillators with different phases are plotted in phase space, the phase points are distributed equally about the trajectory. The points are all moving about the trajectory at the same rate since all the oscillators are identical. This means that the density of phase points on the line is equal everywhere and remains so as a cluster of phase points moves about the trajectory. This is an example of Liouville's theorem that states that the density of points in a phase space of arbitrary dimension remains constant in time

$$d\rho/dt = 0$$

this is easily visualized for the two-dimensional phase space.

13.11 The Langevin Equation

A diatomic molecule with a dipole moment defined by positive and negative charges q separated by a distance d

$$\mu = qd$$

rotates to align with an applied electric field E. This energy to align is opposed by the thermal energy kT to produce a continuous range of alignments from full alignment ($\theta=0$) to full opposition ($\theta=\pi$). The energy at any angle θ is

$$\varepsilon\,(\theta) = -\mu E \cos\,(\theta)$$

Using a differential element $\sin\theta\,d\theta$, the partition function is

$$q = \int_0^\pi \exp\,(+\beta\mu E \cos\,(\theta))\sin\,(\theta)\,d\theta$$

With $x = \cos(\theta)$, $dx = +\sin(\theta)d\theta$, $a = \beta\mu E$, $x_i = \cos(0) = 1$, and $\cos(\pi) = +1$, the integral becomes

$$q = \int_{-a}^1 \exp\,(ax)\,dx = \frac{1}{a}\exp\,(ax)\,|_{-1}^2 = \frac{1}{a}\left[e^{ax} - e^{-ax}\right]$$

The energy for this system $<\varepsilon>$ also determines the average dipole moment in the field since

$$<\varepsilon> = \mu E < \cos\,(\theta) > = <\mu > E$$

The average energy is

$$<\varepsilon> = -\frac{\partial \ln\left[\frac{1}{\beta\mu E}\right]}{\partial\beta}\Bigg] - \frac{\partial \ln[e^{\beta\mu E} - e^{-\beta\mu E}]}{\partial\beta} = +\frac{1}{\beta} - \mu E\frac{e^{\beta\mu E} + e^{-\beta\mu E}}{e^{\beta\mu E} - e^{\beta\mu E}}$$

$$= \frac{1}{\beta} - \mu E \coth\,(\beta\mu E)$$

The average alignment angle is

$$<\cos\,(\theta)> = -<\varepsilon>/\mu E = \coth\,(\beta\mu E) - 1/\beta\mu E$$

Problems

13.1 The three-dimensional kinetic energy distribution function is

$$\frac{dn}{n} = \frac{2}{\sqrt{\pi}}\left(\frac{E}{kT}\right)^{\frac{1}{2}}e^{-\frac{E}{kT}}\frac{dE}{kT}$$

Set up the equation for the average kinetic energy of the gas.
Note:

$$I\,(n) = \int_0^\infty x^{n-1}e^{-x}dx = \sqrt{\pi}$$

$$I\,(n+1) = nI\,(n)$$

13.2 Molecules are free to move on a two-dimensional surface at a temperature of 200 K.

a. Set up the integral for the mean square speed $<v^2>$ for this system.
b. Evaluate the integral(s) for the mean square speed.

13.3 The energies of vibration, E, for a reacting molecule range from 0 to ∞

a. Determine the partition function.
b. Determine the probability that the molecule has vibrational energy between E_0 and ∞.
c. Set up the integral(s) to determine $<E^2>$ for this system.
d. Set up a differential expression that gives $<E^2>$ from the partition function.

13.4 A particle moves continuously on a ring of constant radius with an energy

$$\frac{p^2}{2I}$$

where p is the angular momentum and I is the moment of inertia

a. Determine the partition function for this system by integrating over all possible angular momenta $(-\infty < p < \infty)$.
b. Determine the average energy.

13.5 The probability of finding a total energy E in a molecule containing s identical harmonic oscillators (molecular vibrations) is

$$p\,(E)\,dE = \frac{e^{-\beta E}E^{s-1}dE}{(kT)^s}$$

a. Determine the partition function for this system.
b. Set up an integral expression for the probability of finding a molecule with energies between E_1 and E_2.
c. Set up an integral expression for the average energy.
d. Find the average energy.

13.6 Given that particles of mass m at height h (ground level is $h = 0$) have energies mgh:

a. Set up the integral for the partition function and evaluate it.
b. Determine the average height of the particles.

Chapter 14
Interactions

14.1 A Two-Site Enzyme

A protein with two independent ligand binding sites resolves into two single site partition functions even if the two sites have different binding parameters. The distinct sites can even be on different proteins. Interaction between two sites with bound ions requires an additional interaction energy so that the total partition function cannot be factored. For example, charged ions on two neighboring sites repel each other and the repulsion energy Boltzmann factor is included in the Boltzmann factor for the state with ions at each site.

The partition function for an enzyme with two independent, different binding sites

$$\xi = (1 + q_{b1}\lambda_1)(1 + q_{b2}\lambda_2) = 1 + q_{b1}\lambda_1 + q_{b2}\lambda_2 + q_{b1}\lambda_1 q_{b2}\lambda_2$$

requires an additional Boltzmann factor in the final term if the ligands repel (or attract) each other

$$q_i = \exp(-\beta\varepsilon_i)$$

The partition function

$$\xi = 1 + q_{b1}\lambda_1 + q_{b2}\lambda_2 + q_{b1}\lambda_1 q_{b2}\lambda_{b2} q_i$$

cannot be factored.

The fraction of enzymes with two bound substrates decreases for a large repulsion energy since q_i is small to reduce the fraction.

14.2 Koshland–Nemethy–Filmer Model

Biological systems often involve groups of interacting proteins where the geometry of the group determines the total interaction energy for each state. A functional unit is composed of three identical proteins might have a linear or planar (triangular) geometry (Fig. 14.1).

M.E. Starzak, *Energy and Entropy*, DOI 10.1007/978-0-387-77823-5_14,
© Springer Science+Business Media, LLC 2010

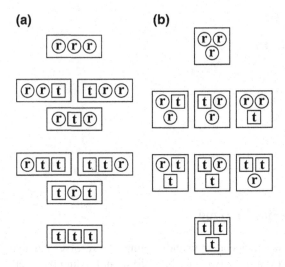

Fig. 14.1 Interaction configurations for three identical interacting proteins: (a) linear and (b) triangular

The linear model has 1–2 and 2–3 interactions, while the triangular model has 1–2, 2–3, and 3–1 interactions. A system of four identical proteins might be linear with three (1–2, 2–3, 3–4) interactions, square planar with four (1–2, 2–3, 3–4, 4–1) interactions or tetrahedral with six (1–2, 1–3, 1–4, 2–3, 2–4, 3–4) interactions (Fig. 14.2).

The Koshland model for protein conformational changes allows each protein to change conformation when it binds substrate

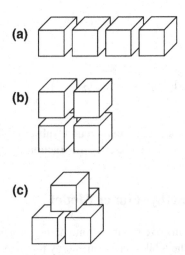

Fig. 14.2 Linear (a), square (b), and tetrahedral (c) interacting configurations for a four protein oligomer

$$r \rightarrow tS$$

r is the reference state. Each independent t conformer requires a Boltzmann factor for the conformational change ($k = \exp(-\beta \Delta g)$), a Boltzmann factor for the energy of binding (q_b) and a chemical potential with Boltzmann factor $\lambda = \exp(\beta \mu)$

$$L = kq_b\lambda$$

A two protein system with no interactions has a partition function

$$\xi = 1 + 2L + L^2 = (1 + L)^2$$

where the degeneracy is 2 for the rt and tr conformations.

A linear three protein system with rt and tt interactions (ε_{rt}, $q_i = \exp(-\beta \varepsilon_{rt})$, $\varepsilon_{tt}, q_{tt} = \exp(-\beta \varepsilon_{tt})$ and $q_{rr}=1$) requires additional Boltzmann factors in the partition function for independent proteins

$$\xi = (1 + L)^3 = 1 + 3L + 3L^2 + L^3$$

The rrt conformers have three orders (rrt, rtr, and trr). rrt is different from trr since the two interacting sides of each protein are different. rrt and trr have one rt interaction, while the middle has two rt interactions. The interactions break the degeneracy

$$3L \rightarrow 2Lq_{rt} + Lq_{rt}^2$$

The $3L^2$ term has arrangements rtt, trt, and ttr. trt has two rt interactions, while rtt and ttr have one tr and one tt interaction

$$3L^2 \rightarrow L^2q_{rt}^2 + 2Lq_{rt}q_{tt}$$

The final linear conformer (ttt) has two tt interactions

$$L3 \rightarrow L^3q_{tt}^2$$

The partition function with rt and tt interactions

$$\xi = 1 + 2Lq_{rt} + Lq_{rt}^2 + L^2q_{rt}^2 + 2L^2q_{rt} + L^3q_{tt}^2$$

determines the number of bound substrates and the thermodynamic parameters of the system. A change in substrate concentration changes the Boltzmann factors for the t conformers so the probabilities of t conformers and the number of bound S change (sigmoidally) as the S concentration increases. The change is more dramatic if t–t interactions are large.

14.3 Surface Double Layers: Ion Surface Interactions

A planar surface with ionizable groups loses its counterions leaving a surface array of fixed surface charges with surface charge density $\sigma = q/A$ – the charges are assumed smeared into a continuum. These fixed ions attract hydrated counterions from the solution while repelling co-ions to produce a double layer. Negative charges on the surface, for example, are balanced by an equal number of positive ions in solution held a distance d from the surface by the waters of hydration. A unit area of negative surface charge balances a unit area of positive solution ions. These two plates of ions of opposite charge separated by distance d define a capacitance in two forms. The separated charges produce an electrical double layer potential, V_d, and capacitance

$$C = q/V_d$$

The capacitance of two plates of area A separated by d is

$$C = \varepsilon A/d$$

where the dielectric constant for the solution $\varepsilon = \varepsilon_0 \varepsilon_r$. ε_r, the relative dielectric constant, is about 80 for water. Equating the two definitions

$$q/V_d = \varepsilon A/d$$

gives a voltage proportional to the surface charge "density" $\sigma = q/A$, the charge per unit area,

$$V_d = dq/\varepsilon A = d\sigma/\varepsilon$$

More realistically, the ions in solution are not all at a fixed distance d from the surface. Because the water and ions in solution are in continuous thermal motion, the counterions in solution are found in a series of aqueous planes increasingly distant from the surface. Thermal energy moves the ions randomly while the electrostatic attraction moves counterions toward the surface. The excess charge in each layer progressively cancels the electric field lines from the negative charge on the surface. At some distance from the surface, the total excess positive charge in all the solution planes completely cancels the negative electric charge emanating from the surface as field lines. At larger distances, the solution is electroneutral. Excess positive charge is largest closest to the surface where the attractive electrostatic energy has its largest negative value, i.e., the electrostatic energy depends on distance from the surface.

The excess charges at each distance x determine the double layer potential $V(x)$ at that location. The surface generates an electric field that diminishes as the field moves through regions with charge of opposite polarity and charge density $\rho(x)$

$$dE/dx = \rho(x)/\varepsilon$$

Since

$$E = -dV(x)/dx$$
$$d^2\,V(x)/dx^2 = -\rho(x)/\varepsilon$$

The excess counterion concentration and charge density depend on the potential via Boltzmann statistics. The voltage $V(x)$ at a distance x from the surface has energy $z_+eV(x)$ and the concentration (for negative V and negative surface charge) is increased relative to the bulk concentration $c_+(b)$:

$$C_+(x) = c_+(b)\exp\left(-\beta z_+eV(x)\right)$$

with a net positive charge per unit area in solution

$$z_+ec_+(x) = z_+ec_+(b)\exp\left(-\beta z_+eV(x)\right)$$

The negative ionic charge at x is

$$z_-ec_-(x) = z_-ec_-(b)\exp\left((-\beta z_-eV(x)\right)$$

The positive and negative concentrations in bulk solution (large x) where the solution is electroneutral are

$$z_+c_+(b) = z_-c_-(b)$$

The net charge density is

$$\rho_{nq}(x) = z_+ec_+(b)\exp\left((-\beta z_+eV(x)\right) + z_-ec_-(b)\exp\left(-\beta z_-eV(x)\right)$$

The second term is negated by the negative z_-. If the exponential arguments are small

$$\exp\left(-x\right) = 1 - x$$

and the net (excess) density is

$$z_+e\,c_+^0\left[1 - \beta z_+eV(x)\right] + z_-e\,c_-^0\left[1 - \beta z_-eV(x)\right]$$

Since the bulk solution is electroneutral

$$z_+c_+(b) + z_-c_-(b) = 0$$

the excess charge density is

$$-\beta eV\left[c_+z_+^2 + c_-z_-^2\right] = -2\beta eVI$$

where the ionic strength

$$I = \frac{1}{2}\left[c_+ z_+^2 + c_- z_-^2\right]$$

equals the salt concentration for a 1:1 electrolyte. Ions with larger charge contribute more to the ionic strength and excess charge density.

The Poisson–Boltzmann equation is

$$d^2 V/dx^2 = -\rho_{nq}(x)/4\pi\varepsilon = +(2\beta eI/\varepsilon)\,V = \kappa^2\,V$$

κ has units m^{-1} and is called the inverse Debye length. The applicable solution is

$$V(x) = A\exp(-\kappa x)$$

The sum of all excess charge at all x facing a unit area of membrane must have the charge per unit area σ. Since

$$\rho(x)/\varepsilon = \kappa^2\,V(x)$$

$$\rho(x) = 4\pi\varepsilon\kappa^2\,V(x) = 4\pi\varepsilon\kappa^2 A\exp(-\kappa x)$$

and, integrating from a, the distance of closest approach to infinity

$$\sigma = \int_a^\infty 4\pi\varepsilon\kappa A\exp(-\kappa x) = 4\pi\varepsilon\kappa^2 A\left(-\frac{1}{\kappa}\right)\exp(-\kappa x)\,|_a^\infty$$

$$= +4\pi\varepsilon\kappa A\exp(-\kappa a)$$

$$A = \sigma\exp(\kappa a)/4\pi\varepsilon\kappa = \left[\sigma\kappa^{-1}/\varepsilon\right]\exp(\kappa a)$$

The potential $V(x)$ is

$$V(x) = \left(\sigma\kappa^{-1}/\varepsilon\right)\exp(-\kappa\,[x-a])$$

At $x=a$, the distance of closest approach, the potential

$$V(a) = \sigma\kappa^{-1}/4\pi\varepsilon$$

compares with the basic single layer model at distance d

$$V = \sigma d/4\pi\varepsilon$$

The inverse Debye length, κ^{-1}, an average of all the excess charge in solution at all x determines the potential at closest approach.

Because the ionic strength I is proportional to the square of the ionic charge, poly-valent ions play a more significant role in reducing the solution range of the excess solution charge. κ^{-1}, the average distance the diffuse layer extends into solution is reduced more effectively with increasing ionic charge.

14.4 The Debye–Hückel Theory

The chemical potential for an ion

$$\mu = kT \ln c$$

is based on independent ions, i.e., no attraction between cations and anions. However, Coulomb attractions are long range $(r)^{-1}$ and a cation in solution attracts an ion atmosphere of excess counterions. The interaction between the central ion and this ion atmosphere is used to determine the activity coefficient that corrects the concentration expression ($RT\ln c$) for chemical potential.

The excess charge now changes as a function of radius r and the appropriate differential equation in three dimensions depends only on r

$$\frac{1}{r^2}\frac{d}{dr}\left(r^2\frac{dV(r)}{dr}\right) = +\kappa^2\, V(r)$$

The appropriate solution

$$V = A\exp\left(-\kappa r\right)/r$$

with

$$A = ze/\varepsilon$$

is verified by substitution. ze is the total central ion charge.

For dilute solutions, the double layer potential at the ion surface of radius R becomes

$$V = \frac{ze\exp\left(-\kappa r\right)}{\varepsilon}\frac{1}{r} = \frac{ze}{\varepsilon R}\frac{1}{\exp\left(\kappa R\right)} = \frac{ze}{\varepsilon R}\frac{1}{1+\kappa R} = \frac{ze}{\varepsilon R}\frac{\kappa^{-1}}{\kappa^{-1}+R}$$

The product of radial terms

$$\frac{\kappa^{-1}}{R\left(R+\kappa^{-1}\right)} = \frac{1}{R} - \frac{1}{R+\kappa^{-1}}$$

Separates the potential into two parts

$$V\left(R\right) = \frac{ze}{\varepsilon R} - \frac{ze}{\varepsilon\left(R+\kappa^{-1}\right)}$$

The first term is the central ion's potential at the surface. The second term is a correction for the ion atmosphere at a distance $R+\kappa^{-1}$ from the center that corrects the potential (and the chemical potential) for the central ion.

14.5 One-Dimensional Ising Model

An electron has two spin states that have different energies in a magnetic field. Some atoms such as iron have an intrinsic magnetic moment that gives them aligned and opposed orientation in a magnetic field. If the atoms aligned with the field exceed those opposed, the metal is magnetized.

A one-dimensional line of such atoms with no interactions obeys two state Boltzmann statistics. For a magnetic moment μ, the aligned ($\varepsilon_+ = -\mu H$, $q_+ = \exp(+\beta\mu H)$) and opposed ($\varepsilon_- = +\mu H$, $q_- = \exp(-\beta\mu H)$) energies give probabilities

$$p_+ = q_+ / (q_+ + q_-) \quad p_- = q_- / (q_+ + q_-)$$

for each atom. For a line of N atoms,

$$\Xi = (q_+ + q_-)^N$$

If $p_+ = 0.75$, 3/4 of the atoms are aligned but, without interactions, these aligned atoms appear randomly along the line. Interactions make the aligned atoms "condense" in a region of the line so that, for example, the left three quarters of the line is strongly magnetized.

When interactions are allowed, the partition function for N atoms is expanded and the interactions for each configuration are calculated. A three atom line has eight (2^3) configurations or terms that can have different interaction energies. Although Boltzmann products can be developed for chains of two or three atoms, the partition function for larger N is calculated using special techniques.

A 2×2 matrix includes all possible Boltzmann factors for a single unit interacting with its neighbor to the right. An atom with a "+" alignment has energy ε_+ $q_+ = \exp(-\beta\varepsilon_+)$. If it interacts with a second "+" atom to its right with interaction energy ε_{++}, $q_{++} = \exp(-\beta\varepsilon_{++})$. The four Boltzmann factors generated in this manner ($q_+ q_{++}$, $q_+ q_{+-}$, $q_- q_{-+}$, $q_- q_{--}$) are all the factors for a single unit and are arranged in a matrix array where the rows tabulate the state of the atom and the columns tabulate the state of its nearest neighbor. The transfer matrix

$i \backslash i+1$	$+$	$-$
$+$	$q_+ q_{++}$	$q_+ q_{+-}$
$-$	$q_- q_{-+}$	$q_- q_{--}$

for one unit is extended to a chain of N atoms as

$$\Xi = M^N$$

if the Nth unit in the chain interacts with the first unit (cyclic boundary conditions). Each atom then sees an interacting unit to its right.

This matrix formulation for a chain is used to produce a modified partition function for each atom in the chain using eigenvalue–eigenvector techniques.

14.6 Eigenvalue Techniques

A matrix is an array of numbers that acts on a vector to rotate and alter its length. The 2×2 matrix

$$\begin{pmatrix} 1 & 1 \\ 1 & 1 \end{pmatrix}$$

acts on the two-component vector $(x,y) = (2,1)$ to give a new vector

$$\begin{pmatrix} 1 & 1 \\ 1 & 1 \end{pmatrix} \begin{pmatrix} 2 \\ 1 \end{pmatrix} = \begin{pmatrix} 3 \\ 3 \end{pmatrix}$$

The original vector with 63.4° and length $\sqrt{2^2 + 1^2} = \sqrt{5}$ rotates to 45° with a new length $\sqrt{3^2 + 3^2} = \sqrt{18}$. The matrix operates on only two "eigenvectors" in the two-dimensional space to change the length but not the orientation of the vector

$$\begin{pmatrix} 1 & 1 \\ 1 & 1 \end{pmatrix} \begin{pmatrix} 1 \\ 1 \end{pmatrix} = 2 \begin{pmatrix} 1 \\ 1 \end{pmatrix}$$

$$\begin{pmatrix} 1 & 1 \\ 1 & 1 \end{pmatrix} \begin{pmatrix} 1 \\ -1 \end{pmatrix} = 0 \begin{pmatrix} 1 \\ -1 \end{pmatrix}$$

The scalar factors of the unrotated eigenvectors are the eigenvalues.

If the eigenvectors are used for the coordinate system, the matrix is transformed for that new coordinate system. The new matrix is diagonal with the two eigenvalues as the diagonal elements

$$\begin{pmatrix} 2 & 0 \\ 0 & 0 \end{pmatrix}$$

Any power of this diagonal matrix is a power of each of the diagonal eigenvalues

$$M_d^N = \begin{pmatrix} 2^N & 0 \\ 0 & 0^N \end{pmatrix}$$

These eigenvalues λ are determined with a characteristic determinant and polynomial

$$\begin{vmatrix} 1 - \lambda & 1 \\ 1 & 1 - \lambda \end{vmatrix} = 0$$

$$(1 - \lambda)(1 - \lambda) - 1 = 0 = \lambda^2 - 2\lambda$$

$$\lambda = 0 \qquad \lambda = 2$$

14.7 An Eigenvalue Partition Function

The general transfer matrix for atomic magnets with nearest neighbor interactions

$$\begin{pmatrix} q_+q_{++} & q_+q_{+-} \\ q_-q_{-+} & q_-q_{-+} \end{pmatrix}$$

is simplified with the following assumptions:

(1) the opposed state is reference ($q_- = 1$);
(2) aligned–opposed interaction energy is 0 ($q_{+-} = q_{-+} = 1$);
(3) aligned–aligned and opposed–opposed have the same interaction energy ($q_{++} = q_{--} = q'$); and
(4) the transfer matrix is

$$\begin{pmatrix} qq' & q \\ 1 & q' \end{pmatrix}$$

The characteristic determinant and polynomial

$$\begin{vmatrix} qq' - \lambda & q \\ 1 & q' - \lambda \end{vmatrix} = 0$$

$$(qq' - \lambda)(q' - \lambda) - q = 0$$

$$qq'^2 - q - q'(1+q)\lambda + \lambda^2 = 0$$

give eigenvalues

$$\lambda_\pm = \frac{+q'(1+q)}{2} \pm \frac{\sqrt{q'^2 - 2qq'^2 + q'^2q^2 + 4q}}{2}$$

Since N is usually large, the larger eigenvalue raised to the Nth power dominates and the partition function is

$$\Xi = \lambda_+^N$$

for cyclic boundary conditions where the Nth atom interacts with the first atom. The single atom partition function is

$$\xi \approx \lambda_+$$

The complicated expression for the atoms with interactions simplifies when the interaction energy is 0 and $q' = 1$

$$\lambda_+ = \frac{1+q}{2} + \frac{\sqrt{1 + 2q + q^2}}{2} = 1 + q$$

the partition function for independent atoms.

14.8 The One-Dimensional Ideal Lattice Gas

A gas condenses to liquid when interactions become important at low volume and temperature. A one-dimensional lattice gas is a line of M cells that are either empty or occupied by one of N gas atoms ($N<M$). The Boltzmann probability that a site is occupied depends on the chemical potential:

$$\lambda = \exp(\beta\mu)$$

N particles on M sites are arranged in

$$M!N!(M-N)!$$

ways to give a partition function

$$Q = \frac{M!}{N!(M-N)!}q^N$$

for an ideal gas with no interactions. q is the partition function for one gas molecule.

The number of sites M is the one-dimensional equivalent of volume. The three-dimensional classical equation

$$P = \left(\frac{\partial A}{\partial V}\right) = -P$$

and the free energy

$$<a> = -kT\ln(Q)$$

suggest the definition for a one-dimensional system pressure

$$P = -\left(\frac{\partial <a>}{\partial M}\right) = +kT\left(\frac{\partial \ln(Q)}{\partial M}\right) = +kT\left[\ln(M) - \ln(M-N)\right]$$

$$= -kT\ln\left(1 - \frac{N}{M}\right) = kT(1-\theta)$$

where $\theta = N/M$ is the "concentration," the number of atoms per total sites. Since

$$Ln(1-\theta) = -\theta$$

for a dilute gas ($N<<M$),

$$P = +kT\theta = kTN/M$$

$$PM = NkT$$

This is the equation of state, analogous to $PV = nRT$, for a one-dimensional ideal gas.

14.9 The Bragg–Williams Approximation

The number of ways N non-interacting particles can be arranged on M sites

$$\frac{M!}{N!(M-N)!}$$

is valid for any geometrical arrangement of the M cells. The equation can be used for a one-, two-, or three-dimensional gas. Interactions between the molecules require a separation of states that have interacting neighbors and those that do not.

The Bragg–Williams approximation uses the fraction or probability that a particle has interacting neighbors to determine its average energy of interaction. Two nearest neighbor particles interact with energy ε and if the system geometry can produce up to c interacting neighbors, the energy of interaction can range from 0 to $c\varepsilon$. The average energy of interaction for one particle is

$$c\varepsilon N/M$$

As the particles fill lattice sites ($N \rightarrow M$), the interactions approach their maximum energy ($c\varepsilon$). Each particle experiences this average interaction so the total interaction energy for the system is

$$N(c\varepsilon N)/2M = c\varepsilon N^2 M$$

The energy is divided by 2 since 1–2 and 2–1 interactions are the same.
The partition function for the lattice gas with M sites and N particles is

$$Q = \frac{M! q^N \, \exp\left(-\beta c\varepsilon N^2/M\right)}{N!(M-N)!}$$

The pressure P for this gas depends only on logarithmic terms with M

$$P + kT\left(\frac{\partial \ln Q}{\partial M}\right) = +kT\frac{\partial}{\partial M}\left[\ln M! - \ln(M-N)! - \beta c\varepsilon N^2/2M\right]$$

$$= +kT\left[\ln M - \ln(M-N) + \beta c\varepsilon N^2/M^2\right]$$

$$P = kT\theta + \beta\varepsilon\theta\, c^2/2$$

This is an expansion in $\theta = N/M$, the "concentration."
The derivative for the chemical potential for this gas acts only on $\ln(N)$ terms

$$\mu = \left(\frac{\partial <A>}{\partial N}\right)_{M,T} = -kT\frac{\partial \ln (Q)}{\partial N}$$

$$= -kT\frac{\partial}{\partial N}\left[-\ln (N!) - \ln (M-N)! + N\ln q - \beta c\varepsilon\frac{N^2}{2M}\right]$$

$$= kT\ln\left[\frac{N}{M-N}\right] - kT \ln q + 2c\varepsilon\frac{N}{M}$$

Attractive interactions ($\varepsilon<0$) produce a lower chemical potential. Particles tend to sites that have interacting neighbors. The first term is the "adsorption" isotherm

$$N/(M-N) = \theta p/(1-\theta)$$

Problems

14.1 A one-dimensional alloy with metals A ($\lambda_A = \exp(\beta\mu_A)$) and B ($\lambda_B = \exp(\beta\mu_B)$) has only A–A ($\varepsilon_{AA}$) and B–B interactions ($\varepsilon_{BB}$). Give the transfer matrix for one site in this alloy.

14.2 The Debye length for a 0.01 M solution of 1:1 electrolyte is $\kappa^{-1} = 3 \times 10^{-9}$ m = 3 nm. Noting that the Debye length is inversely proportional to the square root of the ionic strength, determine the Debye length for 0.01 M $CaCl_2$ and 0.01 M $CaSO_4$.

14.3 The three proteins of an interacting array have two conformations (r,t) and are arranged with a planar, triangular geometry. If the free energy for a change r = t for each conformer is g ($L=\exp(-\beta g)$) and the only interaction energy is t–t (ε_{tt}), determine the partition function and the average number of

(a) t conformers and
(b) ttt oligomers.

Chapter 15
Statistical Thermodynamics in Chemical Kinetics

15.1 The Dog-Flea Model Revisited

The dog-flea model illustrates an equilibrium distribution with the maximal states. If all fleas are on dog A, each might jump to dog B with some characteristic time. Independent jumps by fleas in either direction eventually lead to an equilibrium state. Even then, random jumps create a dynamic equilibrium. Even though fleas continue to jump between dogs, the 50–50 distribution is stable. Fleas continue to jump and although the system can fluctuate from this equilibrium, the 50–50 equilibrium is observed predominantly. For equal energy isomers, an initial state of only A isomers leads, in time, to the 50–50 A–B distribution.

The isomer reaction rate depends on initial conditions. For a non-equilibrium distribution of 16 fleas and 0 fleas on dogs A and B, respectively, the forward rate from A to B is large. Any one of the 16 fleas can jump to dog B. The rate (number of fleas jumping per unit time) is proportional to the number of fleas $N = 16$, on dog A

$$\text{Rate (A)} = -k_1 N_A$$

where the negative signifies loss of fleas from dog A. k is a proportionality constant with units of inverse time, e.g., s^{-1}. Its inverse, the natural lifetime τ, establishes the time scale. Since the departing fleas go to dog B, the rate of gain for B is

$$\text{Rate (B)} = +k_1 N_A$$

Since dog B has no fleas initially, the rate for a jump from B to A is 0

$$\text{Rate (B)} = -k_1 N_B = k_{-1}(0) = 0$$

As fleas jump to dog 2, the forward rate decreases. If $k = 1/4\,s^{-1}$,τ, the natural lifetime $= 2.5\,s$,

$$\text{Rate} = -1/4 N_B = 1/4(16) = -4$$

M.E. Starzak, *Energy and Entropy*, DOI 10.1007/978-0-387-77823-5_15,
© Springer Science+Business Media, LLC 2010

i.e., four fleas have jumped leaving $16-4 = 12$ fleas behind. In the next time interval, the forward rate of loss is smaller since rate depends on N_A at a this time

$$\text{Rate} = -kN1 = 1/4(12) = -3$$

Any of the four fleas on dog B might also return to dog 1 in this time interval. When all fleas (or isomers) have the same energy, $k_1 = k_{-1} = k$, and the rate of return is

$$kN_B = \frac{1}{4}(4) = 1$$

The net rate for A is

$$dN_A/dt = -kN_A + kN_B = -1/4(12) + 1/4(4) = -2$$

As B increases, the overall net rate slows.

For first-order kinetics, isomer transition probabilities are defined for each time; the total time the particle has been a specific isomer is irrelevant. The fraction at a given time is proportional to the number of this isomer at that time.

For a reversible reaction, the rate constant for the decay to equilibrium is the sum of the forward and reverse rate constants

The net rate for B is

$$\frac{dN_A}{dt} = -kN_A + kN_B$$

Using the conservation of total isomers

$$N_A + N_B = N$$

the rate equation is

$$dN_A/dt = -kN_A + k(N - N_A) = kN - (2\,k)N_A$$

The solution for $N_A = N$, $N_B = 0$ at $t = 0$

$$N_A = N\exp(-2kt) + Nk/2\,k[1 - \exp(-2kt)]$$

illustrates an interesting characteristic of first-order reactions. The N_A isomers at the start of the reaction might be expected to decay with rate constant k since no B are present. However, the net rate constant $(2k)$ is the sum of forward and reverse rate constants even though no B is present. The decay reflects transitions, not the specifics of A to B or B to A.

15.2 Reversible Reactions and Equilibrium

The net rate of reaction for the equal energy isomers

$$dN_A/dt = -kN_A + kN_B$$

with conservation of total isomers

$$N = N_A + N_B$$

slows as the system approaches an equilibrium where the net rate is 0

$$0 = -kN_A + k(N - N_A)$$

and

$$kN = 2kN_A$$
$$N_A = N/2$$

Although transitions continue, the numbers of A and B isomers are constant at this dynamic equilibrium. Different $N/2$ molecules are A isomers at any time in dynamic equilibrium.

Transition rate constants are different if the A and B isomers have different free energies. The equilibrium probabilities are now Boltzmann probabilities

$$p_A = \exp(0/kT)/q \qquad p_B = \exp(-g/kT)/q$$
$$q = 1 + \exp(-g/kT)$$

and the forward and reverse rate constants are modified to produce this result at equilibrium. Substituting the Boltzmann probabilities for dynamic equilibrium

$$0 = -k(1/q) + k\exp(-g/kT)/q = -k\,p_A + kr p_B$$

requires rate constants

$$K_1 = k\exp(-g/kT)$$
$$k_{-1} = k$$

These rate constants now define equal forward and reverse rates at dynamic equilibrium. This is the condition of detailed balance at equilibrium.

15.3 Kinetic Averages

Continuous first-order differential equations are solved for $N(t)$. The irreversible kinetic equation

$$dN/dt = -kN$$

with N_0 at $t=0$ gives

$$dN/N = -kdt$$

$$\int_{N_o}^{N(t)} dN/N = \int_0^t -kdt$$

$$\mathrm{Ln}(N/N_0) = -kt$$

$$N(t) = N_0 \exp(-kt)$$

$$P(t) = N_0/N = \exp(-kt)$$

$N(t)$ and $p(t)$ are the average number of particles and probabilities, respectively for time t.

The average time for the decay is determined from the probability

$$p(t) = \exp(-kt)/\int_0^\infty \exp(-kt)dt$$

as

$$\int_0^\infty (t)\exp(-kt)dt/\int_0^\infty \exp(-kt)dt$$

The numerator is the derivative of the denominator with respect to k and the ratio is

$$= -\partial \left[\int \exp(-kt)dt\right]/\partial k/\int_0^\infty \exp(-kt)dt$$

Since

$$\int_0^\infty \exp(-kt)dt = -k^{-1}\exp(-kt)|_0^\infty = 1/k$$

the numerator is

$$d/dk\left[\int_0^\infty \exp(-kt)dt\right] = \int_0^\infty t\exp(-kt)dt = d(k^{-1})/dk = -k^{-2}$$

and the average time is

$$-d(k^{-1})/dk/k^{-1} = +k^{-2}/k^{-1} = 1/k = \tau$$

The natural lifetime, the probabilistic time for a change, is equal to the average time for decay.

The reversible rate expression

$$\frac{dN_A}{dt} = -k_1 N_A + k_{-1} N_B$$

with conservation of particles

$$N_t = N_A + N_B$$

is

$$dN_A/dt = -k_1 N_A + k_{-1}(N_t - N_A) = k_{-1}N_t - (k_1 + k_{-1})N_A$$

The integrating factor $\exp(kt)$ with $k = k_1 + k_{-1}$ is multiplied on each side

$$\exp(kt)[dN_A/dt + kN_A] = k_{-1}N_t \exp(kt)$$

to reduce the left to a single differential

$$\left[N_1 \exp(kt) \right] / dt = k_{-1}N_t \exp(kt)$$

Integrating gives

$$N_A(t) = N_t \exp(-kt) + \frac{k_{-1}}{k} N(1 - \exp(-kt))$$

and

$$N_B(t) = N_t - N_A(t)$$

15.4 Stochastic Theory for First-Order Decay

$N(t)$ for the irreversible first-order irreversible decay

$$N(t) = N_o \exp(-kt)$$

is the average $<N(t)>$. For a given system, the actual value might vary but it remains close to this average. Stochastic analysis is used to determine both the average and the standard deviation (fluctuations) about this average at each time t.

$p(N,t)$ is the probability the system has N molecules at time t. For the next interval Δt, the probability of having N molecules, $P(N, t+\Delta t)$, is determined by two events. (1) There are $N+1$ molecules and one reacts in the time interval leaving N molecules and (2) there are N particles at t and nothing happens in the interval for this irreversible reaction.

The rate constant k is a probability factor for a change in unit time. The product $k\Delta t$ is the probability one particle will react. For $N+1$ particles, the probability one reacts is $k(N+1)\Delta t$. The probability of no reaction for the N particles in the interval is

$$1 - kN\Delta t$$

The total probability of "change" in the interval Δt is

$$P(N, t + \Delta t) = P(N + 1, t)(k(N + 1)\Delta t + P(N, t)(1 - kN\Delta t)$$

$P(N, t+\Delta t)$ is expanded in a Taylor series

$$P(N,t) + (\partial P/\partial t)\, dt = P(N+1,t)k(N+1)dt + P(N,t) - P(N,t)kNdt$$
$$\partial P(N,t)/\partial t = p(N+1,t)k(N+1) - P(N,t)kN$$

This differential equation for $P(N,t)$ is solved using a generating function with a dummy variable s to create a series in the probabilities

$$F(s,t) = \sum P(N,t)s^n$$

Derivatives of this generating function generate averages. For example, the average value of x as a function of time is determined by differentiating the generating function with respect to s and then setting $s = 1$,

$$s\,(\partial F/\partial s)_{s=1} = s\sum P(N,t)Ns^{N-1} = \sum NP(N,t) = <N>$$

The average square deviation from the average value as a function of time is

$$<(N - <N>)^2> = \left(\partial^2 F/\partial s^2\right)_{s=1} + (\partial F/\partial s)_{s=1} - (\partial F/\partial s)^2_{s=1}$$

These average values are found by converting the differential equation in probabilities to a differential equation in the generating function

$$\sum (\partial P(N,t)s^n/\partial t) = k\sum P(N+1,t)(N+1)s^{N+1} - k\sum P(N)Ns^N$$
$$= (\partial F/\partial t) = k(1-s)\,(\partial F/\partial s)$$

The solution to this equation

$$F(s,t) = [1 + (s-1)\exp(-kt)]^{N_0}$$

is verified by substitution.

The average number of reactant molecules at each t is

$$<N(t)> = s\partial F/\partial s|_{s=1}\, [1 + (s-1)\exp(-kt)]^{N_0}$$
$$= N_0\, [1 + (s-1)\exp(-kt)]^{N_0-1}\, |_{s=1}\exp(-kt)$$
$$= N_0\exp(-kt)$$

The average change in reactant A determined by the stochastic analysis is equivalent to that determined from the general rate equation. The stochastic approach also gives the mean square fluctuations from this average value at each t

$$<(N - <N>)^2> = N_0\exp(-kt)[1 - \exp(-kt)]$$

The largest fluctuations are observed at intermediate times for this irreversible reaction.

15.5 The Wind-Tree Model

Boltzmann's statistical approach was criticized because it seemed to violate some fundamental physical laws. Systems move to equilibrium in positive time. However, collisions between particles that produce reaction are symmetric in time. Since Newton's force equation

$$md^2x/dt^2 = F$$

is indifferent to the direction of time. If time is reversed (. t to $-t$), Newton's force equation is unchanged

$$md^2x/d(-t)^2 = md^2x/dt^2 = F$$

How could equations symmetrical in time generate an approach to equilibrium with increasing time?

Ehrenfest proposed a simple model to explain the transition from bidirectional to unidirectional time with a model involving four directional states. An infinite plane is covered randomly with diamond-shaped "trees" with their four vertices facing north (1), east (2), south (3), and west (4), respectively (Fig. 15.1).

"Wind" particles roll along the frictionless plane with constant velocity in only the four possible directions (N, E, S, W). Their direction changes when they strike one face of the tree elastically so all particles maintain the same energy. A particle moving in the north (1) direction would deflect east (2) if it struck the right side of

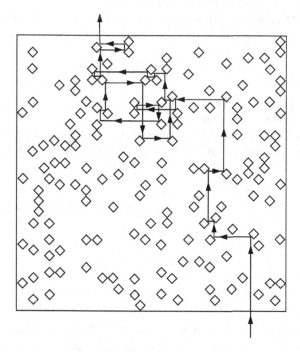

Fig. 15.1 The wind-tree model

the diamond tree; it would deflect west (3) if it struck the left side. A wind particle follows a very definite trajectory through the trees dictated by Newton's laws. If time is reversed, the particle retraces its steps to its starting point. All trajectories are known exactly so entropy is unchanged.

The system becomes time directional when the known trajectories for each particle are replaced by transition probabilities defined by the cross-section length of one tree over the length on the plane allocated to that tree. The probability defines a transition rate constant k for transitions in time. A particle moving north to south changes direction to either west or east. The full set of kinetic equations with north $=1$, north to south $= 3$, west to east $= 2$, and east to west $=4$ is

$$df_1/dt = -2kf_1 + kf_2 + kf_4$$
$$df_2/dt = kf_1 - 2kf_2 + kf_3$$
$$df_3/dt = kf_2 - 2kf_3 + kf_4$$
$$df_4/dt = kf_1 + kf_3 - 2kf_4$$

At equilibrium, each rate equals zero and the equilibrium populations in each direction are

$$f_1 = f_2 = f_3 = f_4 = 0.25$$

The system has reached its most random, highest entropy equilibrium state where all information about the starting populations is lost. Introducing probabilities has produced a unidirectional evolution to equilibrium.

The four rate equations form a rate constant matrix with eigenvalues (decay constants) 0, $-2k$, $-2k$, and $-4k$. $\lambda=0$ is the equilibrium distribution. The solutions are

$$f_i = 0.25 + X_1 e^{-2kt} + X_2 e^{-4kt}$$

The smallest non-zero decay constant (eigenvalue) dictates the evolution to equilibrium (the degenerate $-2k$).

The rate decreases as the system approaches equilibrium and maximal entropy. The stability of the equilibrium distribution is determined from dS/dt for the model

$$S' = -S/K = \sum p_i \ln(p_i)$$
$$= dS'/dt = \sum (dp_i/dt) \ln (p_i) + \sum (p_i/p_i) dp_i/dt = \sum \ln (p_i) dp_i/dt$$

since

$$\sum dp_i/dt = d(1)/dt = 0$$

The kinetic equations for the derivatives are substituted

$$\text{Ln}(p_1)[-2kp_1 + kp_2 + kp_4] + \ln (p_2)[-2kp_2 + kp_1 + kp_3]$$
$$+ \ln (p_3)[-2kp_3 + kp_2 + kp_4] + \ln (p_4)[-2kp_4 + kp_1 + kp_3]$$

Collecting terms in pairs (1,2), (2,3), (3,4), (4,1) gives terms

$$
\begin{aligned}
&\mathrm{Ln}(p_1)(-p_1 + p_2) + \ln(p_2)(-p_2 + p_1) \\
&+ \ln(p_2)(-p2 + p_3) + \ln(p_3)(-p_3 + p_2) \\
&+ \ln(p_3)(-p_3 + p_4) + \ln(p_4)(-p_4 + p_3) \\
&+ \ln(p_4)(-p_4 + p_1) + \ln(p_1)(-p_1 + p_4)
\end{aligned}
$$

Each line of probability pairs can be reduced to the form

$$(p_2 - p_1)\ln[p_1/p_2]$$

For the fractional probabilities, this form is always negative. The derivative

$$-dS'/dt = -d(S/k)$$

is then always positive so that the entropy rate of change with time increases as the probabilities differ. The rate of entropy change at equilibrium (equal probabilities in this case) is zero. The rate is small when the system is close to equilibrium and increases as the system moves further from equilibrium. This is an example of Lyaponouv stability. If the system is perturbed from equilibrium so that its entropy decreases, it returns to equilibrium more rapidly for larger perturbations.

15.6   The Bimolecular Collision Theory

A bimolecular reaction between A and B in the gas phase might occur if these two molecules collide. The rate of A–B collisions constitutes an upper limit on the rate of bimolecular reactions in the gas phase. This rate of collisions between A and B is proportional to the concentrations of A and B

$$Z_{AB} = C[A][B]$$

The proportionality constant C is the rate constant k for a bimolecular reaction where every collision between A and B leads to reaction

$$Rate = \frac{d[A]}{dt} = -k[A]^1[B]^1$$

The proportionality constant C (or k) depends on the relative velocity (the velocity at which they approach each other

$$v_r = \sqrt{\frac{8kT}{\pi \mu}}$$

with reduced mass μ

$$1/\mu = 1/m_A + 1/m_B$$

And the radii r_A and r_B of A and B, respectively.

Z_{AB} is a reaction rate only if each A–B collision produces reaction. Bimolecular collision theory postulates that only some fraction of the molecules collide with enough energy to react. The Boltzmann factor for a collision energy is

$$e^{-\varepsilon/kT} = e^{-\beta\varepsilon}$$

The partition function for the colliding molecules is

$$q = \int_0^\infty e^{-\beta\varepsilon}d\varepsilon = -\frac{1}{\beta}e^{-\beta\varepsilon}\Big|_0^\infty = -\frac{1}{\beta}[e^{-\infty} - e^{(0)}] = \frac{1}{\beta} = kT$$

The probability that the collision has an energy greater than the minimum energy for reaction, ε_m, is found by integrating the probabilities for each energy,

$$p(\varepsilon) = \frac{e^{-\beta\varepsilon}}{q}$$

from the minimum energy for reaction ε_m to ∞,

$$p(\varepsilon < \varepsilon_m) = \int_{\varepsilon_m}^\infty p(\varepsilon)d\varepsilon = \int_{\varepsilon_m}^\infty \frac{e^{-\beta\varepsilon}}{q}d\varepsilon = -\frac{1}{\beta q}e^{-\beta\varepsilon}\Big|_{\varepsilon_m}^\infty = e^{-\beta\varepsilon_m}$$

since $\beta q = \beta\frac{1}{\beta} = 1$

This total probability is the fraction of all collisions that lead to reaction. For A mole of reacting A and B, the probability is

$$p(E > E_a) = e^{(-\varepsilon_a/k_b T)} = e^{(-E_a/RT)}$$

The total number of reactive collisions is

$$\frac{dP}{dt} = Z_{AB}e^{(-E_a/RT)}$$

Comparing this theoretical result

$$\frac{dP}{dt} = Z_{AB}e^{(-E_a/RT)} = Ce^{(-E_a/RT)}[A][B]$$

with the experimental (Arrhenius) rate expression,

$$k_{exp} = C\exp(-E_a/RT)$$

E_a describes the minimum collision energy for reaction and C, calculated from the cross section and relative velocity, describes the frequency factor A.

15.7 Transition State Theory

Non-elastic X–Y collisions can create an XY complex where energy flows through-out the complex. Activated complex theory postulates short-lived complexes that exist for about 10^{-13} s. The XY complex is in pseudo-equilibrium with the X and Y reactants since the activated complexes can dissociate to products.

A bimolecular reaction

$$X + Y \Leftrightarrow XY^{\ddagger} \rightarrow Products$$

has forward and reverse rate constants of k_1 and k_{-1} respectively, and k_2 for the irreversible reaction to products. The dagger notation for the activated complex was a typesetter's error. Eyring had used an asterisk.

Reactants and the transition state complex rapidly reach a pseudo-equilibrium. The reaction to products is ignored to produce an equilibrium

$$X + Y \xrightarrow[k_{-1}]{k_1} XY$$

$$0 = -k_1[X][Y] + k_{-1}[XY^{\ddagger}]$$

$$K^{\ddagger} = \frac{k_1}{k_{-1}} = \frac{[XY^{\ddagger}]}{[X][Y]}$$

$$[XY^{\ddagger}] = K^{\ddagger}[X][Y]$$

The rate of reaction to product is

$$R = k_2[XY^{\ddagger}] = k_2 K^{\ddagger}[X][Y]$$

k_2 is the frequency for a single vibration. The vibration with quantum energy $h\nu$ is thermalized to the classical vibrational energy kT

$$kT = h\nu \qquad \nu = \frac{kT}{h}$$

This frequency is the rate constant for complex dissociation

$$k_2 = \nu = \frac{kT}{h}$$

$$\tau_2 = \frac{1}{\nu} = \frac{h}{kT}$$

At 300 K, the k_2 is approximately 10^{13} s^{-1}. The molecule could dissociate in 10^{-13} s.

The experimental bimolecular rate expression

$$\frac{dP}{dt} = +k[X][Y]$$

is compared with the rate for the transition state theory to give a theoretical expression for the experimental rate constant

$$k = \frac{k_b T}{h} K^{\ddagger}$$

This basic form is valid for all types of reaction. Only the equilibrium constant changes. The unimolecular reaction

$$A \Leftrightarrow A^{\ddagger} \rightarrow P$$

has equilibrium constant

$$K^{\ddagger} = \frac{A^{\ddagger}}{A}$$

$$dP/dt = k_2 A^{\ddagger} = k_2 K^{\ddagger} A$$

15.8 The Energetics of Transition State Theory

The empirical rate constant has an activation energy E_a

$$k_{\exp} = A e^{(-E_a/RT)}$$

The transition state rate constant

$$k = K^{\ddagger} k_2$$

is converted to an energy form using the hypothetical free energy for complex formation

$$0 = \Delta A^{o\ddagger} + RT \ln K^{\ddagger} \qquad \Delta A^{o\ddagger} = -RT \ln K^{\ddagger} \qquad K^{\ddagger} = e^{\frac{\Delta A^{\ddagger}}{RT}}$$

The transition state rate constant

$$k = \frac{k_b T}{h} K^{\ddagger} = \frac{k_b T}{h} e^{\frac{\Delta A^{\ddagger}}{RT}}$$

is modified with

$$\Delta A^{\ddagger} = \Delta E^{\ddagger} - T \Delta S^{\ddagger}$$

$$e^{-\frac{\Delta A^{\ddagger}}{RT}} = e^{-\frac{(\Delta E^{\ddagger} - T \Delta S^{\ddagger})}{RT}} = e^{+\frac{\Delta S^{\ddagger}}{R}} e^{-\frac{\Delta E^{\ddagger}}{RT}}$$

$E^{\ddagger}$ correlates with the experimental activation energy. The temperature independent entropy factor is incorporated into the frequency factor A

$$E_a = \Delta E^{o\ddagger} \qquad A = \frac{kT}{h} e^{(\Delta S^{\ddagger}/R)}$$

$$A = \frac{kT}{h} e^{\frac{\Delta S^{\ddagger}}{R}}$$

since

$$k = Ae^{\frac{E_a}{RT}} = \frac{k_b T}{h} e^{\frac{\Delta S^{o\ddagger}}{R}} e^{\frac{\Delta E^{o\ddagger}}{RT}}$$

The model predicts a frequency factor directly proportional to T.

Both reactant and complex free energies require the same standard state. However, the hypothetical activated complex has no tabulated free energy; it must be estimated.

15.9 Transition State Theory and Partition Functions

The unimolecular reaction with activated complex

$$A \Leftrightarrow A^{\ddagger} \rightarrow P$$

has equilibrium constant

$$K^{\ddagger} = \frac{[A]^{\ddagger}}{[A]}$$

involving the postulated, short-lived activated complex. The concentration of this complex is estimated using partition functions since this "sum over states" sums all possible energy states of a molecule.

The partition functions for reactant molecules are calculated from their known energies. The partition function for the activated complex is found by postulating a structure and estimating its partition function.

The equilibrium constant for the formation of activated complex from molecules A and B

$$K^{\ddagger} = \frac{[AB]^{\ddagger}}{[A][B]}$$

requires an $AB^{\ddagger}$ partition function. Since the partition functions are all based on a common reference state, the free atoms, the equilibrium constant is

$$K^{\ddagger} = \frac{Q_{AB}^{\ddagger}}{Q_A Q_B}$$

$Q_{AB}^{\ddagger}, Q_A$, and Q_B are each products of partition functions for translation, rotation, and vibration. For example,

$$Q_A = q_{At}^i q_{Ar}^j q_{Av}^k$$

where i, j, and k are the number of translational, rotational, and vibrational degrees of freedom, respectively.

The activated complex is expected to have a higher energy than the reactants but conservation of energy requires that it have the same total energy. The difference in the partition function lies in the larger number of states and larger partition function created by the new geometry of the complex.

An accurate estimate of a rate constant using transition state theory requires detailed energy information. Errors cancel for ratios making it possible to make estimates for partition functions in the ratios.

The one-dimensional translational partition function is

$$q_t = \sqrt{\frac{2\pi RT}{M}}$$

N_2 (MW $= 0.028$ kg) at 300 K has a partition function of 748 while He has a partition function of 1978. An estimate of 10^3 establishes the order of magnitude. For translation in three dimensions, the partition function is $q_t{}^3 = 10^9$.

An average rotational partition function $q_r = 10^2$ per rotation. A quantum oscillator has a partition function

$$q_v = 1 + e^{-1\beta\varepsilon} + e^{-2\beta\varepsilon} + \cdots$$

between 1 and 10.

A diatomic reactant molecule (AB) with three translations, two rotations, and one vibration and an estimated partition function

$$Q_{AB} = q_t^3 q_r^2 q_v^2 \approx (10^3)^3 (10^2)^2 (5)^1 = 5 \times 10^{13}$$

reacts with an atom C with partition function

$$Q_c = q_t^3 = (10^3)^3 = 10^9$$

The activated complex might be triangular or linear. A linear complex has three translations, two rotations, and $3(3)-5 = 4$ vibrations. One vibration stretches and breaks with rate constant $v = kT/h$. The remaining three are tabulated in the partition function

$$Q_{AB\ddagger} = q_t^3 q_r^2 q_v^3$$

The equilibrium constant is

$$K^\ddagger = \frac{Q_{AB\ddagger}}{Q_A Q_B} = \frac{q_t^3 q_r^2 q_v^3}{q_t^3 \left[q_t^3 q_r^2 q_v^1 \right]} = \frac{q_v^2}{q_t^3} = \frac{25}{10^9} \approx 25 \times 10^{-9}$$

and the estimated rate constant is

$$k = k_2 K^\ddagger = \frac{k_b T}{h} K^\ddagger = (10^{13})(25 \times 10^{-9}) = 25 \times 10^4$$

15.10 The Lindemann Mechanism

A gas reaction that obeys first-order kinetics, i.e., concentration to the first power, implies a unimolecular reaction. Each molecule reacts independently. A second-order reaction implies a bimolecular reaction where two molecules interact to react. Certain gas phase reactions can display both orders. They are first order at high pressure or concentration but are second order at lower pressures or concentrations. This suggests that the reaction from reactants to products might proceed by two or more steps. The Lindemann mechanism is a two-step sequence. The reactant A molecules are mixed in a heat bath of B molecules which can transfer energy to A on collision to create an excited A^* molecule. This molecule has sufficient energy to react to product although a second collision with a B removes this energy (the strong collision assumption) and returns the molecule to its non-reactive state

The three kinetics rates are (1) activation, (2) deactivation, and (3) reaction to products

$$A + B \rightarrow A^* + B$$
$$A^* + B \rightarrow A + B$$
$$A^* \rightarrow P$$

have rate constants k_1, k_{-1}, and k_2, respectively.

The rate for A^* is

$$\frac{d[A^*]}{dt} = +k_1[A][B] - k_{-1}[A^*][B] - k_2[A^*]$$

The steady state approximation,

$$\frac{dA^*}{dt} = 0$$

states that the small concentration or pressure of A^* formed is constant

$$0 = k_1[A][B] - k_{-1}[A^*][B] - k_2[A^*]$$
$$k_1[A][B] = [A^*](k_{-1}[B] + k_2)$$
$$[A^*] = \frac{k_1[A][B]}{k_{-1}[B] + k_2}$$

The rate of formation of products is

$$\frac{dP}{dt} = k_2[A^*] = k_2\frac{k_1[A][B]}{k_{-1}[B] + k_2}$$

This equation reduces to first or second order under the proper conditions. For large $[B]$, $k_{-1}[B] >>$. k_2 in the denominator is dropped to give a first-order rate of formation

$$\frac{dP}{dt} = \frac{k_2 k_1[A][B]}{k_{-1}[B] + 0} = \frac{k_2 k_1}{k_{-1}}[A] = k^e[A]$$

The limiting first-order rate constant is a combination of rate constants

$$k^e = \frac{k_2 k_1}{k_{-1}}$$

The equilibration of A and A^* at high pressure gives an equilibrium constant

$$K^e = \frac{k_1}{k_{-1}} = \frac{[A^*][B]}{[A][B]} = \frac{[A^*]}{[A]}$$

The rate constant expression has the same form as the activated complex model; both have an initial equilibration step and a reaction step to products

$$k = K^e k_2$$

at high B pressures.

A–B collisions decrease with decreasing B pressure slowing formation and deactivation of A^*. At low pressure, $k_{-1}[B] \ll k_2$ so $k_{-1}[B] \approx 0$

$$\frac{dP}{dt} = \frac{k_2 k_1 [A][B]}{0 + k_2} = k_1 [A][B]$$

The second-order rate depends only on k_1. The slowest, or rate-determining step, is the one that creates the A^*. Once created, the molecule remains active for a time sufficient to react without deactivating.

The Lindemann rate includes intermediate pressures where both first- and second-order processes contribute to the observed rate.

15.11 Bose–Einstein Statistics

The discrete energy packets that add to a quantum mechanical oscillator are photons. Photon statistics deals with integral numbers of indistinguishable photons. This Bose–Einstein statistics differs from Boltzmann statistics because indistinguishability changes the procedure for counting states.

The $2^2 = 4$ Boltzmann states for a system with two molecules and two absorbed photons recognize labeled photons. If the molecules are labeled 1 and 2 and the photons a and b, then the states are all possible because the a and b quanta are distinguishable. For indistinguishable quanta, the second and third states are one state (Fig. 15.2).

The "molecules" can be vibrational modes in a single molecule. If the photons are free to move between vibrational modes, sufficient photons (energy) can collect in one mode to induce a vibrational rupture, i.e., reaction.

A molecule with two equal vibrational modes accepts three photon of the proper energy. Four distinct states are possible: (1) three quanta in mode 1, (2) three quanta

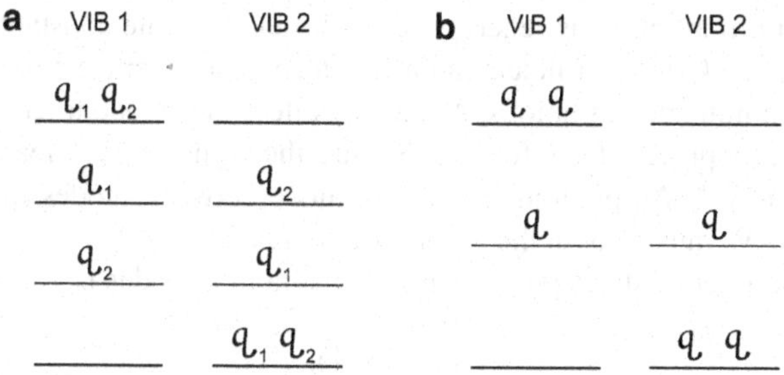

Fig. 15.2 Boltzmann (a) and Bose–Einstein (b) states for two oscillators and two quanta

in mode 2, (3) 2 quanta in mode 1 and 1 in mode 2, and (4) 1 quantum in mode 1 and 2 in mode 2.

Four Bose–Einstein states when the photons are indistinguishable compare with $2^3 = 8$ Boltzmann states.

The number of Bose–Einstein states (Chapter 10) for g modes and N photons is

$$\Omega = (N + g - 1)!/N!(g - 1)!$$

For 2 modes ($g=2$) and 2 photons ($N=2$), the number of states is

$$(2 + 2 - 1)!/(2!)(2 - 1)! = (N + g - 1)!/(N!)(g - 1)! = 3!/2!1! = 3$$

For 2 modes and 3 photons ($N=3. g=2$), the number of states is

$$(3 + 2 - 1)!/3!(2 - 1)! = 4!/3! = 4$$

The total Bose–Einstein states

$$\frac{(N + g - 1)!}{N!(g - 1)!}$$

give the total states possible. For $N=3$. $g=2$ with four total states, two of the states have all three photons in a single vibration. If three vibrational quanta were sufficient to break the bond, two out of four states would lead to reaction to give a reaction probability 2/4=1/2.

Three photons ($N=3$) and three modes ($g=3$) have 10 total states. Three of these 10 have all three photons in one mode; if three photons are sufficient to break the bond, the probability of reaction is 3/10 =0.3.

15.12 Energy Dependence of k_2

The Lindemann mechanism postulates a single excited molecule. However, the B molecules have a distribution of energies and can create A^* with different quanta. An excited A^* with more photons (energy) reacts faster. The rate constant k_2 is energy dependent and the k_2 in the Lindemann mechanism is an average rate constant.

A minimum number of photons, N_o, must collect in a mode to create a reactive A^*. The only acceptable states for reaction are those that have a least N_o photons in the bond. Only $N–N_o$ photons are free to move between modes in A^* to define reactive states; N_o must stay in the bond for reaction.

The total number of states possible for N quanta in g modes is

$$\frac{(N + g - 1)!}{N!(g - 1)!}$$

The number of states possible once N_o photons are locked in a mode is

$$\frac{(N - N_o - g - 1)!}{(N - N_o)!(g - 1)!}$$

This number of states is divided by total states for the probability of reaction that depends on N, the total photons or energy in the molecule.

When $N=4$ and $j =3$, the state total is

$$\frac{(4 + 3 - 1)!}{4!(3 - 1)!} = \frac{(6)!}{4!2!} = \frac{6 \times 5}{1 \times 2} = 15$$

If 3 photons are locked in a bond for reaction ($N_o =3$), the number of states is

$$(4 - 3 + 3 - 1)!/(4 - 3)!(3 - 1)! = 3!/2! = 3$$

The probability of reaction is

$$p(\text{reaction}) = 3/15 = 0.2$$

Adding one more photon ($N=5$) increases the probability of reaction

$$\frac{\frac{(5-3+3-1)!}{(5-3)!(3-1)!}}{\frac{(5+3-1)!}{5!(3-1)!}} = \frac{6}{21} = 0.286$$

The probability for N quanta and g vibrational modes with a minimum number of quanta, N_o, per reactive mode is

$$\frac{(N - N_o + g - 1)!}{(N - N_o)!(g - 1)!}$$

The probability that at least N_o quanta collect in a specific vibrational mode is

$$p(> N_o) = \frac{\frac{(N-N_o+g-1)!}{(N-N_o)!(g-1)!}}{\frac{(N+g-1)!}{N!(g-1)!}}$$

Since the molecule has a distribution of energies, these probabilities are averaged and multiplied by a time scaled constant k to determine an average k_2.

15.13 The Continuum Approximation

The A^* reaction probabilities are summed over all possible N. For large, the probabilities are recast in an energy continuum ($\varepsilon = Nh\nu$). The probability for k_2 is converted to $k_2(\varepsilon)$ using Stirling's approximation,

$$M! = M^M e^{-M}$$

The exponential terms cancel in both numerator and denominator

$$\frac{e^{-(N-N_o+g-1)}}{e^{-(N-N_o)}e^{-(g-1)}} = \exp\left(-N - N_o - g + 1 + N - N_o + g - 1\right) = e^0 = 1$$

to give

$$\frac{\frac{(N-N_o+g-1)^{N-N_o+g-1}}{(N-N_o)^{N-m_o}(g-1)^{j-1}}}{\frac{(N+g-1)^{N+j-1}}{N^N(g-1)^{j-1}}} = \frac{(N-N_o+g-1)^{N-N_o+g-1}N^N}{(N-N_o)^{N-N_o}(N+g-1)^{N+g-1}}$$

If $N \gg g$, $g-1$ is ignored in the parentheses while n is retained in the powers to avoid an oversimplified probability of 1

$$\frac{N - N_0^{N-N_o+g-1}N^N}{N - N_0^{N-N_o}N^{N+g-1}} = \frac{(N-N_o)^{g-1}}{N^{g-1}} = \left(\frac{N-N_o}{N}\right)^{g-1}$$

Total energy in the molecule is

$$E = N\varepsilon = Nh\nu$$

while the minimum energy is

$$E_o = N_o\varepsilon = N_o h\nu$$

$k_2(E)$ is

$$k_2(> E_o) = k\left(\frac{E-E_o}{E}\right)^{g-1}$$

In the limit of high pressure, the Lindemann unimolecular rate constant

$$\frac{k_1}{k_{-1}}k_2 = K_1 k_2$$

becomes a Boltzmann probability times $k_2(E)$ for each energy. For 1 mode, the vibrational Boltzmann probability

$$e^{(-E/k_bT)}\frac{dE}{kT}$$

is used. The partition function for j identical oscillators is

$$(k_bT)^j$$

The Boltzmann probability for g oscillators is developed by analogy with differential elements for multidimensional systems. The radial two-dimensional differential element is

$$rdr$$

while the three-dimensional element is

$$r^2dr$$

A differential element in an M-dimensional radial space is

$$r^{M-1}dr$$

The differential element for g identical oscillators is

$$E^{g-1}dE$$

The probability for an energy E for g oscillators is

$$\exp{(-E/kT)}(E/kT)^{g-1}d(E/kT)$$

The result is also the ratio of forward and reverse rate constants for an excited energy E.

The average rate constant for j modes the integral of the rate constant k_2 and Boltzmann probability

$$<k_2> = \int_{E_o}^{\infty} k(E)p(E)dE = \int_{E_o}^{\infty} k\left(\frac{E-E_o}{E}\right)\left(\frac{E}{kT}\right)^{g-1}e^{-\beta E}dE/kT$$

using the probability distribution for the g indistinguishable oscillators

$$\frac{e^{-\beta E}}{(j-1)!}\left(\frac{E}{kT}\right)^{g-1}d\left(\frac{E}{kT}\right)$$

For intermediate pressures, the average rate is the integral over the Lindemann expression

$$dP/dt = \int_0^\infty \frac{k_2(E)k_1[A][B]}{k_{-1}[B] + k_2(E)} p(E)dE$$

15.14 Energy Transfer

Two molecules with vibrational energy can collide to form a complex where the energy moves though all the vibrational modes of the complex. When the complex separates into its individual molecules, energy can be transferred. When a D molecule with three photons strikes a second molecule B with none to form a complex, what is the probability that has transferred its three quanta to A?

Maximal states occur for the complex since it contains all the photons and all the modes. Information on the initial modes and photons for D and A is lost as energy moves throughout modes in the complex. The number of states for the separate D and A molecules after dissociation is the product of the numbers of states for each. This probability of energy transfer is the ratio of the total states for the separated D and A over the total states for the complex. D has 3 photons ($N=3$) and 2 equivalent modes ($g=2$) initially and A has zero photons ($N=0$) and 3 modes with the same energy ($k=3$) initially. What is the probability that, after transfer, D has zero photons ($N=0$) and A has 3 photons ($M=3$)?

The complex has $M+N = 3$ total photons and $g+k = 2+3 = 5$ equivalent vibrations and

$$(3 + 5 - 1)!/3!(5 - 1)! = 7!/3!4! = 7 \times 6 \times 5/3 \times 2 = 35$$

states.

After separation, D has 0 photons (and 2 modes) for

$$(0 + 2 - 1)!/0!(2 - 1)! = 1$$

state (the energy free state). The A molecule has three photons and three modes for a total of

$$(3 + 3 - 1)!/3!(3 - 1)! = 5!/3!2! = 5 \times 4/2 = 10$$

states. The probability of this energy transfer is

$$p(D(3)A(3)) = 10 \times 1/35 = 2/7$$

If the complex forming collision rate Z_{DA} is known, the product of this rate and the probability of transfer gives the rate of energy transfer.

In this example, the complete transfer of three quanta was the only energy transfer considered in determining the probability. The problem requires averaging when different numbers of photons are transferred on collision.

Problems

15.1 A Lindemann mechanism requires the formation of an excited A^* molecule by collisions with heat bath molecules B. Further collisions with B will remove the energy to return the molecule to the unexcited state A.

a. Determine the steady-state concentration of A^* if it can return to A either by collisions with B (rate constant k_{-1}) or by emitting the excess energy as a photon of radiation (rate constant k_f).
b. Determine the intensity of radiation emitted by the A^* molecules.

15.2 A linear diatomic molecule A decomposes via a unimolecular mechanism. For transition state theory, it proceeds through the intermediate A

$$A \Leftrightarrow A^{\ddagger} \rightarrow P$$

a. Tabulate the degrees of freedom for A and $A^{\ddagger}$, i.e., list number of translations, rotations, etc.
b. If $q_{trans} = 10^3$, $q_{vib} = 10$, and $q_{rot} = 100$, determine $K^{\ddagger}$ for this reaction.
c. Determine the rate constant k for the reaction if $(kT/h) = 10^{13} \times s^{-1}$.
d. Determine $\Delta A^{\ddagger}$ for this reaction at 300 K $(RT = 2500 \text{ J mol}^{-1})$.

15.3 An atom A interacts with a diatomic molecule (BC) to form a triangular activated complex in three-dimensional space. Determine $K^{\ddagger}$ for the equilibrium between reactants and the activated complex using partition functions.

15.4 Two A atoms which exist in the gas phase above a surface can both bind to the surface and interact to form an activated complex on the surface. Because the complex is confined to two dimensions (the surface), it has two possible translations in the plane.

a. Determine the number of rotations and vibrations possible for the activated complex.
b. Set up (do not evaluate) the partition function for the activated complex using q_t, q_{rot}, etc.
c. Give the theoretical rate constant for the reaction in terms of the partition function (do not evaluate).

15.5 A molecule A with N photons and g equivalent vibrational modes collides with a second molecule B with no quanta and k vibrational modes.

a. How many different ways can the N photons be distributed when the two molecules combine to form a complex.
b. What is the probability that the molecule A has only i photons when the complex dissociates, i.e., $N-i$ quanta are transferred to B.

15.6 The rate constant for a bimolecular process is described by the Arrhenius expression

$$k(T) = Ae^{(-E_a/RT)} = 3 \times 10^9 e^{(-50,000/RT)}$$

a. If this rate constant is described with the collision theory model what is the collision frequency.
b. What is the minimal energy required to create a reactive molecule.

Chapter 16
Irreversible Thermodynamics and Transport

16.1 Charge Flux

The product of electrical potential ψ and charge dq is a reversible free energy change

$$dG = \psi dq$$

Charge is transferred without changing the potential. The energy is the product of an intensive $\psi(J/C)$ and extensive (qC) parameters. A battery converts chemical energy to electrical energy. Energy is transferred reversibly only when the current that moves across the system boundary is infinitesimally small.

Batteries are normally used under irreversible conditions. Charge flows as a current and the free energy or work of the battery is transferred to the surroundings. This work is degraded to heat. The energy transferred finally to heat is described by having the surroundings approximated by a resistor. As current flows through the resistor under the battery potential, the resistor produces heat.

The current through a resistor for a given potential difference is defined by Ohm's law, a linear equation; the current $i(C/s)$ is directly proportional to the electrical potential difference ψ

$$i = \psi/R = G\psi$$

The conductance G (Siemens = amperes/volt) is the inverse of resistance in ohms. While classical thermodynamics has no time variable, the current is charge per second for this irreversible system. The potential is also modified for this irreversible system. The distance over which the potential acts determines the current generated. A grounded electrode placed 1 km from 20,000 V produces no current since the conductance of all the intervening air is too low. The same electrode 1 mm from the 20,000 V ionizes the air to produce a large current. The difference behavior is due to the potential gradient (the electric field) in each case

$$E = -d\psi/dx$$

M.E. Starzak, *Energy and Entropy*, DOI 10.1007/978-0-387-77823-5_16,
© Springer Science+Business Media, LLC 2010

The electric field is

$$20,000/0.001 = 2 \times 10^7 \, \text{V/m}$$

for the 1 mm separation and 20 V/m for the 1 km separation. The minus sign provides consistent flow charge. If the positive potential on the right is larger, it makes a positive charge move to the left – the electric field vector direction.

The conductance of the material increases with the distance between electrodes. To provide a consistent, general Ohm's law, a current "density" or more accurately a charge flux is defined as the current per unit area

$$J_q = i/A(C/s/m^2)$$

For a homogeneous material of length d

$$E = -\frac{\psi}{d}$$

Combining these intensive quantities in Ohm's law gives

$$J_q = i/A = -[Gd/A]\psi/d = [Gd/A]E = +\kappa E$$

$Gd/A = \kappa$ is the conductivity, i.e., the conductance across a 1 m long cube with 1 m^2 cross-section. The sign is positive since a positive field moves a positive test charge from left to right.

The irreversible current flow involves time and distance in charge flux and electric field, respectively. These independent variables were absent in classical thermodynamics but now characterize irreversibility.

The product of current and voltage is power (J/s^{-1})

$$P = i\psi$$

for a volume $V = Ad$. The power per unit volume

$$\frac{P}{V} = \frac{iV}{Ad} = \frac{i}{A}\frac{\psi}{d} = JE$$

is the product of the flux and driving force. Power is the rate of production of heat. Since

$$q = TS$$

the rate of heat production at constant temperature is related to the rate of entropy production

$$\varphi = \frac{dq}{dt} = T\frac{dS}{dt} = J_q E$$

TdS/dt is the dissipation function.

16.2 Generalized Forces and Fluxes

Ohm's law

$$J_q = \kappa E$$

is one example of a general type of linear equation called a flux force equation. The electric field, a force per unit charge, produces a charge flux. A concentration gradient dc/dx produces a flow of particles in a linear equation

$$J_n = L\frac{dc}{dx}$$

A force X based on a potential gradient produces a flux. The general force flux equation

$$J = LX$$

is common to many systems close to equilibrium. Forces and fluxes are generated from conjugate thermodynamic variables. For example, Ohm's law is created from the conjugate pair $\psi\, dq$ by defining a force as the gradient of the electric potential (the intensive variable)

$$X = E = -d\psi/dx$$

and the flux as the charge (the extensive variable) per unit time and area

$$J_q = \frac{1}{A}\frac{dq}{dt}$$

The linear Ohm's law then relates these new variables

$$J = LX \qquad J_q = \kappa E$$

Force flux equations are derived from any pair of thermodynamic conjugate variables. Each pair (PV, Vq, μn) is the product of an intensive variable (a potential) and an extensive conjugate (charge). A generalized potential, (Y) and a generalized charge Q give a thermodynamic energy

$$YQ$$

The general linear force flux equation has a one-dimensional force using the intensive quantity

$$X = -\frac{dY}{dx}$$

and flux using the extensive conjugate

$$J_Q = \frac{1}{A}\frac{dQ}{dt}$$

The force flux relation is

$$J = \frac{1}{A}\frac{dQ}{dt} = LX = -L\frac{dY}{dx}$$

The thermodynamic pair μn gives a force

$$X = -\frac{d\mu}{dx}$$

and flux

$$J = \frac{1}{A}\frac{dn}{dt}$$

and the linear relation

$$J = \frac{1}{A}\frac{dn}{dt} = LX = -L\frac{d\mu}{dx}$$

For the PV conjugate pair, the pressure gradient,

$$X = -\frac{dP}{dx}$$

is the pressure difference across the piston of width Δx. The flux is

$$J = \frac{1}{A}\frac{dV}{dt}$$

The change in volume is $dV = Adx$ and

$$J = \frac{1}{A}A\frac{dx}{dt} = \frac{dx}{dt} = v$$

The flux is the piston velocity. This flux can be interpreted as the motion of volume elements across the piston. The force flux equation is

$$J = v = +LX = -L\frac{dP}{dx}$$

The dissipation function

$$\Phi = JX = |vdP/dx|$$

shows that both entropy and heat production rates increase with gradient and velocity.

16.3 Particle Flux

The flux of particles is proportional to the gradient of the chemical potential

$$J_n = 1/A\,dn/dt = -L\,d\mu/dx = LX$$

for n moles. The mole flux is also the product of particle velocity v and concentration c

$$J_n = -vc$$

with units $v(m)c$ (moles/m^3) $= vc$ mol/m^2.

The force mobility u' is the velocity per unit force

$$u' = v/X \qquad v = uX$$

Comparing the equations

$$j = -vc = -u'cX = -u'c\,d\mu/dx = -L\,d\mu/dx$$
$$L = u'c$$

A single particle flux is related to the particle concentration c

$$J_N = u'c\,d\mu/dx$$

with single particle chemical potential

$$d = kT\,d\ln(c)$$

The particle flux is

$$J_N = -u'cd[kT\ln(c)]/dx = -u'\frac{ckT}{c}\frac{dc}{dx} = -u'kT\frac{dc}{dx}$$

Comparing this with Fick's empirical first law

$$J = -D\,dc/dx$$

gives the Einstein equation

$$D = u'kT$$

u' has units velocity per force. The frictional coefficient f, defined for force F and velocity v,

$$F = fv$$

has units of force per velocity. The frictional coefficient is the inverse of the force mobility

$$u' = 1/f = v/F$$

The Einstein equation is

$$D = u'kT = kT/f$$

Continuous steady state diffusion occurs across a thin homogeneous region, e.g., a film or membrane, flanked by bathing solutions with concentrations c_1 and c_2, respectively. Since the baths are large, their concentrations remain effectively constant for a flux across the membrane. Since the flux across the region is constant, Fick's law is integrated from bath 1 $(x = 0)$ to bath 2 $(x = L)$

$$\int_0^L J dx = -\int_{c_1}^{c_2} D dc$$

$$JL = -D(c_2 - c_1)$$
$$J = -D/L(c_2 - c_1) = -P(c_2 - c_1)$$

P, the permeability coefficient, has units of velocity (m/s).

16.4 Discrete State Membrane Transport

At the molecular level, particle transport proceeds in a series of discrete jumps where the particle moves between "holes" in the transport medium. Discrete state transport is illustrated with a one dimensional, one site model; particles from either bath jump to a single site in the center of the homogeneous film or membrane. The continuum flux equation

$$J = vc$$

is modified for the three states (bath 1, site, bath 2) of the one site model. $J = vc_1$ is the flux into a unit area of sites. $J = vc_2$ is the flux into the sites from bath 2 and vc^* is the flux from the site into each bath. The stationary state rate of change for c^* is

$$0 = vc_1 + vc_2 - (v + v)c^*$$
$$c^* = [vc_1 + vc_2]/2v = [c_1 + c_2]/2$$

The net flux from the site to bath 2 is

$$J_{net} = vc^* - vc_2 = v/2[c_1 + c_2] - vc_2 = -v/2(c_2 - c_1)$$

The permeability coefficient $P = v/2$. The net flux from bath 1 to site 1

$$J_{net} = vc_1 - vc^* = -v/2(c_2 - c_1)$$

is the same to maintain the stationary state concentration.

For two interior sites, the two stationary state equations for c_1^* and c_2^* are

$$0 = vc_1 + vc_2^* - 2vc_1^*$$
$$0 = vc_1^* + vc_2 - 2vc_2^*$$

The equation is recast as a matrix

$$0 = \mathbf{V}|c^* > +|j >$$

$$0 = \begin{pmatrix} -2 & 1 \\ 1 & -2 \end{pmatrix} \begin{pmatrix} c_1^* \\ c_2^* \end{pmatrix} + \begin{pmatrix} c_1 \\ c_2 \end{pmatrix}$$

$$|C^* > = \mathbf{V}^{-1}|j >$$

$$\begin{pmatrix} c_1^* \\ c_2^* \end{pmatrix} = \frac{1}{3} \begin{pmatrix} 2 & 1 \\ 1 & 2 \end{pmatrix} \begin{pmatrix} c_1 \\ c_2 \end{pmatrix} = \begin{pmatrix} (2c_1 + c_2)/3 \\ (c_1 + 2c_2)/3 \end{pmatrix}$$

The net flux

$$J_n = vc_1 - vc_1^* = -(v/3)(c_2 - c_1)$$

The permeability coefficient is now one third the velocity that the particle moves between sites. For N discrete sites

$$P = v/(N + 1)$$

for a homogeneous film or membrane.

16.5 Fick's Second Law

The stationary state flux described by Fick's first law involves only position x. In irreversible thermodynamics, both x and t are key variables, i.e., the magnitude of the flux changes with time.

A layer of dye solution at $x = 0$ sandwiched between pure water forms a sharp boundary at $t = 0$. With time, the dye moves symmetrically into the water to form a more diffuse layer with a Gaussian shape (Fig. 16.1). $c(x,t)$, the concentration at each x and t is determined using Fick's second law.

Fick's second law is derived from the first law using the equation of continuity in one dimension. Solute passes through a region Δx where some dye remains. The change in flux leaving the region is

$$J_{out} = J_{in} - \left(\frac{\partial J}{\partial x}\right) dx$$

The net particle change for an area A (JA)

$$J_{out}A - J_{in}A = -\left(\frac{\partial J}{\partial x}\right) dxA$$

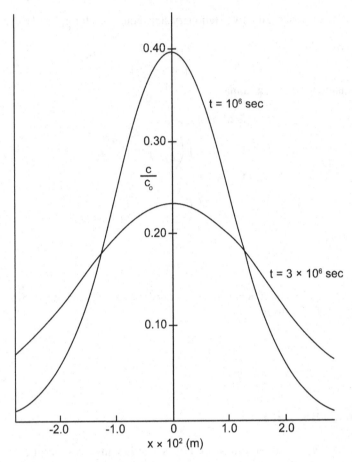

Fig. 16.1 The Gaussian distribution for the concentration $c(x,t)$ with $c = c_0$ only at $x = 0$ at $t = 0$

is equal to the increase in particles within the volume Adx

$$\frac{dc(x,t)}{dt}Adx$$

The two expressions for particle absorption are equated to give the equation of continuity

$$\frac{\partial c}{\partial t}Adx = -\frac{\partial J}{\partial x}Adx$$

$$\frac{\partial c}{\partial t} = -\frac{\partial J}{\partial x}$$

To convert Fick's first law into Fick's second law for $c(x,t)$, the first law is differentiated with respect to x

$$\frac{\partial J}{\partial x} = -D\frac{\partial^2 c(x,t)}{\partial x^2}$$

and the equation of continuity is substituted to give Fick's second law,

$$\frac{\partial c(x,t)}{\partial t} = D\frac{\partial^2 c(x,t)}{\partial x^2}$$

with

$$C(x,t) = c_0 p(x,t)$$

The solution for this partial differential equation if all the material is concentrated at position $x = 0$ at time $t = 0$ is proportional to a Gaussian function with an inverse time argument

$$p(x,t)\alpha \exp(-x^2/4Dt)$$

The probability is formed by a "partition function" an integral of all possible $p(x)$ at any time

$$\int_{-\infty}^{\infty} \exp(-x^2/2Dt)dx = \sqrt{\pi/a} = \sqrt{2\pi Dt}$$

i.e.,

$$p(x,t) = 1/\sqrt{4\pi Dt}\exp(-x^2/4Dt)$$

The average location of the particles is

$$<x> = 1/\sqrt{4\pi Dt}\int_{-\infty}^{\infty} x\exp(-x^2/4Dt)dx = 0$$

since the integral is the product of an odd (x) and even (the exponential) function over the interval.

The squared displacement is

$$<x2> = 1/\sqrt{2\pi Dt}\int_{-\infty}^{\infty} x^2 \exp(-x^2/4Dt)$$

$$= \frac{1}{\sqrt{4\pi Dt}}\sqrt{4\pi Dt}2Dt = 2Dt$$

The root mean square displacement measures the spread of the Gaussian with time.

16.6 Discrete State Diffusion

Diffusion about $x = 0$ proceeds by discrete steps of length d in time intervals τ. The discrete positions are labeled $0, 1, ..., -1, -2$. With no external force, probabilities left and right are equal

$$p_{+1} = p_{-1} = 1/2$$

A transition occurs in each time interval.
For one step, the particle is at -1 or $+1$ with equal probability.
For two steps, the possible steps and final locations are

$$
\begin{aligned}
+1, & \quad +1x = +2 \\
+1, & \quad -1x = 0 \\
-1, & \quad +1x = 0 \\
-1, & \quad -1x = -2
\end{aligned}
$$

The four outcomes for 2τ are three locations. The particle is located at $x = 1$ one out of four times, i.e., $p_2 = 1/4$. $p_0 = 1/2$ since the particle reaches 0 by two paths. At $x = -2$, $p_{-2} = 1/4$. For two steps, the probability for each path is the product $(\frac{1}{2})(\frac{1}{2}) = 1/4$.

After 1τ, the average location of the particle is

$$<x> = 1/2(-1) + 1/2(+1) = 0$$

while its mean square location is

$$<x2> = 1/2(-1)^2 + 1/2(+1)^2 = 1$$

for a root mean deviation of ± 1 about $x = 0$.
After 2τ, the average location is again

$$<x> = 1/4(-2) + 1/2(0) + 1/4(+2) = 0$$

Probabilities in both directions are equal so, on an average, the steps left equal the steps right. The mean square excursion does increase with τ

$$<(x - <x>)^2> = <(x - 0)^2> = 1/4(-2)^2 + 1/2(0)^2 + 1/4(+2)^2 = 2$$

The root mean square distance is

$$X_{rms} = \sqrt{2} = 1.414$$

The binomial coefficients that define the step probabilities are $(1,3,3,1)$ for 3τ. The particle can move to $x = -3$ or $x = +3$ in one way (three steps left or right).

Three paths bring it to +1: (+1,+1, −1), (+1, −1,+1), and (−1,+1,+1). Three paths bring it to −1 as well. These $2^3 = 8$ paths have probabilities

$$p(- 3) = 1/8 = p(+ 3) \qquad P(- 1) = p(+ 1) = 3/8$$

The average location is again 0

$$< x >= 1/8(- 3) + 3/8(- 1) + 3/8(+ 1) + 1/8(+ 3) = 0$$

while the average squared location is

$$< (x - 0)2 >= \frac{1}{8}(9) + \frac{3}{8}(1) + \frac{3}{8}(1) + \frac{1}{8}(9) = \frac{24}{9} = 3$$

For N intervals (steps), the average location is zero and $X_{rms} = \sqrt{N}$ or distance $= \sqrt{N}\,d$

As the size of the steps and the length of time between jumps decrease, the discrete step one-dimensional system evolves to the Gaussian distribution of Fick's second law

$$p(x,t)\alpha \exp (- x^2/2Dt)$$

The diffusion coefficient in terms of d and τ is

$$D =< x^2 > d^2/\tau$$

16.7 The Nernst Planck Equation

The two terms of the electrochemical potential

$$\tilde{\mu} = RT \ln (c) + zF\, \psi$$

or

$$\tilde{\mu} = k_b T \ln (c) + ze\psi$$

give a sum of concentration and electrical forces,

$$X = -\frac{d\tilde{\mu}}{dx} = -\frac{kT}{c}\frac{dc}{dx} - ze\frac{d\psi}{dx}$$

Since the ions carry charge, the flux is either a particle flux with units of mol s^{-1} m^{-2} or a charge flux with units of Cs^{-1} m^{-2}.

The particle flux is converted to a charge flux by multiplying by the charge per particle

$$J_q = zeJ_N$$

The particle flux

$$J = u'cX = vc$$

requires the force X

$$J = cu'\left[-\frac{k_bT}{c}\frac{dc}{dx} - ze\frac{d\psi}{dx}\right] = u'k_bT\frac{dc}{dx} + u'cze\frac{d\psi}{dx}$$

The first term is Fick's first law since

$$D = u'k_bT$$

The second term is Ohm's law for particle flux. For charge flux (current), the conductivity is

$$ze(u'cze) = u'c(ez)^2$$

Ion flow across a film or membrane is driven by both concentration gradient and an electrical potential gradient (electric field). The concentrations and voltages on side 1 and side 2 are $(c_1, 0)$ and (c_2, ψ). The electric field is assumed constant for the homogeneous medium

$$E = -\psi/d$$

and

$$J = -Ddc/dx - u'zeEc$$
$$J/D = -dc/dx - Bc \quad B = u'zeE/D$$

is solved with the integrating factor

$$\exp(+Bx)$$
$$-[J/D]\exp(Bx) = d/dx[c\ \exp(Bx)]$$
$$\int_0^{Ld} -J/D\ \ \exp(Bx)dx = \int_{c_1}^{c_2} d[c\ \exp(Bx)]$$
$$-J/DB\ [\exp(Bd) - 1] = c_2\exp(Bd) - c_1$$
$$J = -DB(c_2\exp(Bd) - c_1)/(\exp(Bd) - 1)$$

Since $Bd = ze\psi/kT$ and $DB = D/d(ze\psi/kT)$, the Nernst Planck constant field equation (the Goldman equation) is

$$J = -D/d(ze\psi/kT)[(c_1 - c_2 \exp(ze\psi/kT))/(1 - \exp(ze\psi/kT))]$$

For large positive values of ψ (potential on side 2 much greater than the potential on side 1), the exponential terms dominate and the equation is

$$J_N = D/L(ze\psi/kT)c_2 = u'zec_2E$$

and

$$J_q = zeJ_N = (ze)^2u'c_2E$$

Comparing this with Ohm's law

$$J_q = \kappa E$$

gives the conductivity for this system.

$$\kappa = (ze)^2u'c_2$$

If the electrical mobility (velocity per electric field)

$$u = zeu'$$

is used,

$$\kappa = zeuc_2$$

and the equivalent conductivity for the ion (conductivity/concentration) is

$$\lambda = \kappa/c = zeu$$

if $\psi_1 \gg \psi_2$, the conductivity depends on the concentration in bath 1

$$\kappa = zeuc_1$$

For $\psi = 0$, $E = 0$. The denominator is zero and J is undefined. However, the limit as V approaches zero is finite and determined using the expansion

$$\exp(x) = 1 + x$$

$$J = -\frac{D}{d}\frac{ze\psi}{kT}\left\{\frac{c_1 - c_2 e^{\frac{ze\psi}{kt}}}{1 - e^{\frac{ze\psi}{kt}}}\right\}$$

$$= -\frac{D}{d}\frac{ze\psi}{kt}\frac{c_1 - c_2 - c_2\frac{ze\psi}{kt}}{1 - 1 - \frac{ze\psi}{kt}}$$

$$= +\frac{D}{d}\left(c_1 - c_2 - c_2\frac{ze\psi}{kr}\right)$$

As $\psi = 0$,

$$J = D/L[c_1 - c_2] = -\frac{D}{L}(c_2 - c_1)$$

16.8 Discrete Diffusion with Drift

The Nernst–Planck equation with constant flux is converted to an equation in x and t using the equation of continuity

$$\left(\frac{\partial J}{\partial x}\right) = -D\left(\frac{\partial^2 c}{\partial x^2}\right) - zeu\left(\frac{\partial cE}{\partial x}\right) = -\left(\frac{\partial c}{\partial t}\right)$$

This second-order equation describes diffusion as well as drift. Particles at $x = 0$ move randomly about $x = 0$ but also drift along the axis. This is illustrated using the discrete state formulation.

Drift from $x = 0$ in a preferred direction results from an applied vector force like electric field. For the discrete step model, this applied force favors one probability. For pure diffusion, $p_1 = p_r = \frac{1}{2}$. If $p_r > p_1$, the particle drifts to the right and the average location changes in time.

For 1τ, the particle reaches $+1$ with $p_r = 0.75$ and -1 with $p_1 = 0.25$. The average location is

$$< x > = 0.75(+1) + 0.25(-1) = +0.5$$

The particle has drifted right with velocity $d/2\tau$.
The square displacement is determined about the average location

$$< x^2 > = 1/4(-1 - 0.5)^2 + 3/4(+1 - 0.5)^2 = 3/4$$

The square deviation is smaller; the directed motion has "ordered" the random motion.

For two steps, the single paths to $+2$ or -2 have joint probabilities

$$(3/4)(3/4) = 9/16 \quad and \quad (1/4)(1/4) = 1/16$$

respectively. The two paths to 0 have joint probability

$$(1/4)(3/4)$$

The average location is

$$< x > = (1/4)(1/4)(-2) + 2(1/4)(3/4)(0) + (3/4)(3/4)(+2) = 1d$$

The particle drifts twice as far as it did for one step, i.e., constant drift velocity. The square displacement about the average location

$$< (x - < x >)^2 > = 1/16(-2 - 1)^2 + 2(3/16)(0 - 1)^2 + 9/16(2 - 1)^2$$
$$= 9/16 + 6/16 + 9/16 = 1.5$$

is also smaller than the deviation for random, equal probability steps.

For three steps, the binomial coefficients (1:3:3:1) describe the paths. Three steps left (1 path) occurs with joint probability

$$(1/4)(1/4)(1/4) = 1/64$$

The three paths to -1 each have two steps left and one right for a joint probability

$$(1/4)(1/4)(3/4) = 3/64$$

and 9/64 for the three paths

The probability for +1 and +3 are both 27/64. The probabilities sum to 1 (1+9+27+27 = 64).

The probability for 2 steps right and 1 left

$$\frac{3!}{2!1!} p_r^2 p_l^1$$

is generalized for N total steps with n steps right and $N{-}n$ steps left

$$\frac{N!}{n!(N-n)!} p_l^{N-n} p_r^n$$

For N total steps (or time intervals) with n jumps to the right and $N-n$ jumps to the left, the particle is located at position

$$x = n - (N - n) = 2n - N$$

with probability

$$\frac{N!}{n!(N-n)!} p_r^n p_l^{N-n}$$

The average location is

$$<2n - N> =! \sum_{n=0}^{N} (2n - N) \frac{N!}{n!(N-n)!} p_r^n p_l^{N-n}$$

16.9 Force Coupling

The Nernst–Planck equation

$$J_n = -uc\frac{d\mu}{dx} - ucze\frac{d\psi}{dx}$$

has two separate force terms to produce the particle flux. The particle force gives a particle flux. However, the electric field also gives a particle flow since the particle moves with its charge. The particle flux is written as

$$J_1 = L_{11}\frac{d\mu}{dx} + L_{12}\frac{d\psi}{dx}$$

The same two forces also produce a charge flux J_2

$$I = J_2 = zeJ_1 = zeucRT\frac{d\mu}{dx} + (ze)^2 uc\frac{d\psi}{dx}$$

$$= L_{21}\frac{d\mu}{dx} + L_{22}\frac{d\psi}{dx} =$$

The two equations

$$J_1 = L_{11}X_1 + L_{12}X_2$$
$$J_2 = L_21X_1 + L_{22}X_2$$

have equal cross-coupling coefficients

$$L_{12} = ucze = L_{21}$$

The is one example of Onsager's reciprocal relations

$$L_{ij} = L_{ji}$$

For example, in the coupled set of equations

$$J_1 = L_{11}X_1 + L_{12}X_2$$
$$J_2 = L_{21}X_1 + L_{22}X_2$$
$$L_{12} = L_{21}$$

16.10 Streaming Current and Electroosmosis

An applied pressure gradient produces a flux of water through a membrane. An applied electric field produces a flow of charge as ions in the same solution. These direct flux–force relations are joined by cross-coupling relations. The applied pressure produces an ion flow (the streaming current). The applied electric field produces a water flux (electroosmosis). The water and charge fluxes (using P for dP/dx) are

$$J_w = L_{wp}P + L_{wE}E$$
$$J_q = L_{qp}P + L_{qE}E$$

L_{qE} is the conductivity κ.

When pressure moves a solution, both cations and anions are carried equally with the water flow and the net flow cannot produce the separation of charge necessary to produce a potential or current.

If the electrolytic solution moves in channels whose walls are lined with negative fixed charge, a excess of positive ions are found near the surface where the solution flows more slowly. The negative solution ions, repelled from the surface, are located in faster moving solution to produce a net charge separation and potential called the streaming potential.

The Onsager reciprocal relations do hold when the units for force are consistent. Pressure in atmospheres differs from electric field (force per charge). However, the electrical mobility, the velocity per unit electric field is developed for water flow past a charged surface. The flow also produces a streaming current and a cross-coupling force–flux relation

$$J_q = L_{qp}P$$

The complementary cross-coupling equation results when an electric field (or potential) drives ions past a surface with fixed charge. Water also flows by the surface, a phenomenon called electroosmosis.

A surface with fixed negative charge per unit area σ and area A has a total charge σA. An applied electric field E parallel to the surface produces a force

$$\sigma AE$$

on the mobile solution plate with an equal number of excess counterions. This motion is opposed by a frictional force which depends on solution viscosity coefficient η, area A, and gradient v/d where v is the velocity of the solution and d is the separation between the fixed charges in the surface and the solution counterion "plate."

$$\sigma AE = \eta Av/d$$

The mobility is the velocity of the solution plate divided by the field

$$v/E = u = \sigma d/\eta$$

This gives either the velocity of the counterions to give a streaming current or the velocity of the water to give the volume flow in accord with Onsager's reciprocal relations.

16.11 Saxen's Relations

The Onsager reciprocal relations connect diverse forces and fluxes. These Saxen's relations act like Maxwell's relations but are valid only for linear relationships between forces and fluxes. For the general coupled force flux relations

$$J_1 = L_{11}X_1 + L_{12}X_2$$
$$J_2 = L_{21}X_1 + L_{22}X_2$$

L_{12} and L_{21} are expressed as ratios of the forces and fluxes and then equated using Onsager's reciprocal relation $L_{12} = L_{21}$.

If $X_1 = 0$ in J_1 equation and $X_2 = 0$ in J_2 equation

$$J_1 = L_{12}X_2$$
$$J_2 = L_{21}X_1$$

so that

$$L_{12} = (J_1/X_2)_{X_1}$$
$$L_{21} = (J_2/X_1)_{X_2}$$

Since $L_{12} = L_{21}$

$$(J_1/X_2)_{X_2} = (J_2/X_1)_{X_1}$$

If $J_1 = 0$ in equation 1

$$0 = L_{11}X_1 + L_{12}X_2$$
$$L_{12}/L_{11} = -(X_1/X_2)_{J_1}$$

If $X_2 = 0$, the ratio of the two equations is

$$(J_2/J_1)_{X_2} = (L_{21}X_1)/(L_{11}X_1) = L_{21}/L_{11}$$

Onsager's relationship gives

$$(J_2/J_1)_{X_2} = -(X_1/X_2)_{J_1}$$

The last ratio uses $J_2 = 0$

$$J_2 = 0 = L_{21}X_1 + L_{22}X_2$$

to give

$$(X_2/X_1)_{J_1} = -(L_{21}/L_{22})$$

and $X_1 = 0$ to form the ratio

$$(J_1/J_2)_{X_1} = L_{12}/L_{22}$$

so that

$$(J_1/J_2)_{X_1} = -(X_2/X_1)_{J_1}$$

Saxen's relations are used to determine an unknown force or flux by controlling some experimental variables (the subscripts) while measuring the remaining three. Force and flux ratios are determined if only two of the four variables are known.

16.12 Scalar Forces and Fluxes

Forces and fluxes are predominantly vector quantities. The force establishes a direction for the flux. A chemical reaction also describes a flow from reactant to product but has no direction in space. The reaction proceeds everywhere in the reaction vessel. This scalar flux from reactant to product is the reaction rate. The gas phase reaction

$$A \rightleftharpoons B$$

has product formation rate (the rate of advancement ξ)

$$R = +dA/dt = \xi = +k_1A - k_{-1}B$$

proportional to the A and B concentrations.

The driving force for reaction is free energy. When $\Delta G = 0$, the reaction is at equilibrium with zero rate (reaction flux). A negative free energy for some pressures A and B means a net forward reaction to equilibrium. For a linear relationship, a larger free energy produces a faster rate of reaction for chemical systems very close to equilibrium

$$R = L\Delta G$$

The Curie symmetry principal states that forces of different order cannot couple. A scalar force, e.g., free energy, cannot produce a vector flow of A or B in homogeneous solution. The Curie principal, however, holds only in homogeneous media. A membrane system can be used to couple scalar and vector processes.

A protein P is completely insoluble in a membrane and cannot be transported across the membrane. If this protein reacts with ATP to form a membrane soluble complex P^*, the reaction will produce a concentration of P^* in the interfacial (side 1) region of the membrane. This concentration produces a vector flow across the membrane. The "scalar" chemical reaction has produced a vector flow. Of course, the reaction is not really homogeneous; it occurs only at one membrane interface. The overall result is a coupling of a reaction and a vector flow.

The kinetic sequence with concentration P_1^* of protein on side 1

$$P + ATP \underset{k_{-1}}{\overset{k_1}{\rightleftharpoons}} P_1^* \overset{k}{\longrightarrow} P_2^*$$

has a stationary state rate of formation for P_1^* and its transfer from side 1 to Side 2. The stationary state fraction of P^*

$$k_1[P][ATP] - (k_{-1} + k)P_1^* = 0$$
$$P_1^* = \frac{k_1[ATP][P]}{k_{-1} + k}$$

gives a net flux to side 2

$$J = \frac{k_1 k[ATP][P]}{k_{-1} + k}$$

The vector flux increases with ATP concentration.

Problems

16.1. The hydrostatic pressure P produces a flow of water across a membrane. The osmotic pressure $\pi = cRT$ where c is solute concentration produces a solute flux through this same membrane.

Set up the two coupled equations for water and solute flow for this system. Discuss the coupling for this system.

16.2. A thermocouple produces a voltage (or current) when its ends are at different temperatures. Set up the pair of force flux equations for this system.

16.3. Show that the square deviation for a one dimensional walk with $p_1 = p_r = 0.5$ is 4 after four steps.

16.4. Set up the forces and fluxes for system with variables T and P from the energies for these two thermodynamic variables.

Chapter 17
Stationary State Thermodynamics

17.1 Introduction

Classical thermodynamics characterizes macroscopic systems at equilibrium; statistical thermodynamics generates averages of thermodynamic parameters from microscopic molecular information. Irreversible thermodynamics explores macroscopic systems displaced, but close to equilibrium. However, the majority of systems are neither at or near equilibrium. Far from equilibrium, new phenomena require a more global examination of the roles of energy and organization. Far from equilibrium, a system might settle into a meta-stable equilibrium maintained by a flow of energy or particles. Far from equilibrium, a homogeneous system might spontaneously order. Reactive systems oscillate through metastable states without coming to a single stable equilibrium. Aspects of these more complicated phenomena are introduced using some simple models.

Irreversible thermodynamics focuses on heat or entropy production. Although entropy increases for the whole universe during an irreversible process, this entropy production can be apportioned so that the entropy of the system decreases while the entropy of the surroundings increase more to produce a net increase for the universe. Schrodinger in "What is Life" noted that the organization and entropy decrease observed for living organisms could be explained if the entropy of their surroundings, e.g., food digested, increased more to increase the entropy of the universe. Of course, the food itself must have a lower entropy before digestion and this lower entropy is produced by the sun through photosynthesis. This progression could continue. The sun's entropy is increasing as it radiates energy. What lowered its entropy in the first place? However, flow systems are the focus of this chapter and it is necessary to distinguish between the sun as an ideal machine – a work generating Carnot cycle between the hot sun and the cold earth or an energy flow system.

Equilibrium statistical mechanics is used to calculate the probability that a living system could come into existence. The appropriate Boltzmann factors are generated from the energy required for the bonds and structure of the organism. Even for a minimal "living" organism such as a virus, this probability is vanishingly small. An organism cannot spontaneously arise in an equilibrium medium even if the appropriate starting materials such as those on the primeval earth are present.

M.E. Starzak, *Energy and Entropy*, DOI 10.1007/978-0-387-77823-5_17,
© Springer Science+Business Media, LLC 2010

The situation changes if the earth is viewed as part of a flow system. Energy from the sun flows through the biosphere to the earth. This unidirectional flow with its intrinsic directional order has the ability to reduce entropy in the biosphere with a concomitant increase of entropy for the sun–earth universe.

Energy flow can produce ordering in homogeneous systems. A liquid such as water heated from below to produce an upward energy flows produces turbulent behavior as the liquid boils. However, if a heavy oil, e.g., spermaceti, is carefully heated from below, the liquid will "spontaneously" organize. The liquid rises in a stable column and then falls through other stable columns – a spatial oscillation. Viewed from above, the rising columns outline hexagonal cells; the falling liquid columns are the centers of these cells (Fig. 17.1). This structure is stable while the steady flow of heat through the system is maintained.

A vortex whistle seems to defy the second law of thermodynamics. Air at one temperature is forced into tube through a lateral inlet and hot air exits one end of the tube while cold air exits the opposite end. The tube creates a temperature gradient. The key to the separation in the shape of the input nozzle which forces the linear stream of air into a vortex within the tube. Since the entire stream now rotates at a constant angular speed, the "hotter", i.e., faster moving, molecules move to the outside, i.e., larger radii, while the cooler molecules circulate at the smaller radii. The outlets are designed to collect, respectively, the gases at the larger and smaller radii.

A bunsen burner uses a flow of gas and air to produce a cone of burning material. The shape of this single cone is ordered but the degree of order is enhanced by careful adjustment of the flow. With the right flow and mixture conditions, the stable single cone separates into two, four, or eight minicones that remain symmetrical and stable while the gas flow is maintained.

A salt solution on a watch glass normally evaporates so that the salt is deposited homogeneously on the glass surface. In some cases, the salt is deposited in a series of concentric Liesegang rings. As water evaporates, the solution becomes supersaturated. When the supersaturation cannot be maintained, the salt precipitates to form a ring. This solution continues to evaporate and supersaturate until it, too, becomes unstable. The salt precipitates to form the next concentric salt ring. The process continues until all the water has evaporated.

The mercury beating heart is an oscillating system. Mercury in a watch glass is covered with an oxidizing solution that produces ions that bind to the mercury surface. The surface charge produces a change in surface tension that causes the mercury surface to expand. A metal electrode is placed in the solution near, but not in contact, with the mercury. As the reaction and expansion proceed, the mercury touches the electrode and discharges its surface charge. The mercury pool returns to its original size and, as the reaction continues, expands again to touch the electrode so that the drop expands and contracts like a beating heart.

Although the mercury drop expands and contracts in time, the chemical reaction that drives it proceeds at a steady pace, i.e., new ionic charge is generated continuously. Some chemical reaction oscillates with time. Reactant forms product but this product is not at final equilibrium; the system remakes reactant to start the process

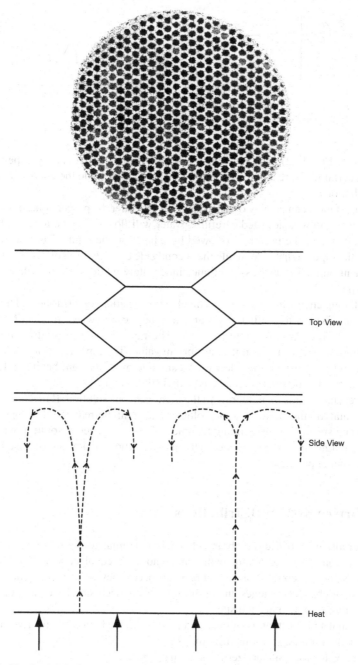

Fig. 17.1 The Rayleigh–Bernard phenomenon

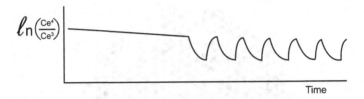

Fig. 17.2   Time course of an oscillating (autocatalytic) reaction

once again (Fig. 17.2). The oscillations do not occur without energy expenditure; some material in the reaction mixture is consumed to provide the energy to sustain the oscillations.

A leaky faucet can display oscillatory behavior under the proper conditions. If the leaking water flow is adjusted carefully, water will flow in bursts; a burst of water on the left side of the faucet is followed by a burst on the right. The flow of water empties the pipe on that side, while the second side continues to collect water for its subsequent burst. The process continues indefinitely while water is added steadily to the pipe.

Oscillating chemical reactions can display both spatial organization and temporal oscillation. The key to oscillations is an autocatalytic step where product formation triggers the production of more products. The rapid growth in product moves its concentration well away from its equilibrium value. The rapid reaction depletes the reactant so the excess product decays. Meanwhile more reactant begins to form so that the autocatalytic process can be repeated (Fig. 17.2).

The reactant builds up randomly in the homogeneous solution. If reactant reaches the autocatalytic threshold in some volume, autocatalytic reaction develops in that volume and spreads to surrounding volumes. The homogeneous solution will order with expanding regions of product. The autocatalytic reaction and its oscillations produce spatial patterns.

17.2 Driven System Distributions

An ensemble of two site membrane channels is similar to the dog-flea or isomer models if ions from adjoining baths can bind with equal energy. If both baths have the same ion concentration and channel occupancy is limited to one site, the ensemble of channels reaches an equilibrium where each site has equal occupancy probability, i.e., $p=0.5$ at each site.

The equal and constant input rates j add ions from the bath while the rate constant k is the same for all transitions from the sites

The rates for each site are zero at stationary state

$$dS_1/dt = j - 2kS_1 + kS_2$$
$$dS_2/dt = j - 2kS_2 + kS_1$$

The two equations in matrix form

$$0 = |j> +\mathbf{K}|S>$$

$$0 = \begin{pmatrix} j \\ j \end{pmatrix} + \begin{pmatrix} -2k & k \\ k & -2k \end{pmatrix} \begin{pmatrix} S1 \\ S2 \end{pmatrix}$$

has a solution

$$|S> = -\mathbf{K}^{-1}|j>$$

with

$$K^{-1} = \frac{-1}{3k} \begin{pmatrix} 2 & 1 \\ 1 & 2 \end{pmatrix}$$

to give site populations

$$\begin{pmatrix} S_1 \\ S_2 \end{pmatrix} = \frac{1}{3k} \begin{pmatrix} 2 & 1 \\ 1 & 2 \end{pmatrix} \begin{pmatrix} j \\ j \end{pmatrix} = \begin{pmatrix} j/k \\ j/k \end{pmatrix}$$

The populations of each state are equal when the inputs from both baths are equal; the populations depend on the time independent ratio j/k. Fractional populations are determined by dividing by this ratio.

The system changes from a dynamic equilibrium to a stationary state system if the two input rates differ; different bath concentrations or an applied voltage produce different input fluxes j_1 and j_2.

The formal solution for the two-site system is the same (Fig. 17.3)

$$|S> = -K^{-1}|j>$$

but

$$|j> = \begin{pmatrix} j_1 \\ j_2 \end{pmatrix}$$

and the stationary state concentrations are different

$$\begin{pmatrix} S_1 \\ S_2 \end{pmatrix} = \frac{-1}{3k} \begin{pmatrix} 2 & 1 \\ 1 & 2 \end{pmatrix} \begin{pmatrix} j_1 \\ j_2 \end{pmatrix} = \frac{1}{3k} \begin{pmatrix} 2j_1 + j_2 \\ j_1 + 2j_2 \end{pmatrix}$$

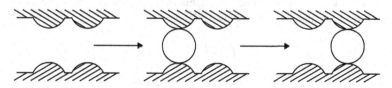

Fig. 17.3 The channel drawing

If $j_2=0$, the absolute populations at stationary state are $S_1 = \frac{2j}{k}$ and $S_2 = \frac{j}{k}$ and the fractional populations are $f_1=2/3$ and $f_2= 1/3$.

If $j_1 \neq j_2$, the total population is the sum of both site populations,

$$j/k = \frac{(3j_1 + 3j_2)}{3k} = \frac{(j_1 + j_2)}{k}$$

The fractional populations are

$$f_1 = \frac{2j_1 + j_2}{3(j_1 + j_2)}$$

$$f_2 = \frac{j_1 + 2j_2}{3(j_1 + j_2)}$$

These stable populations are maintained with a steady flux through the channel. For $j_2=0$, the net flux out of the channel is

$$j_n = k(S_2) = k(j_1/3k) = j_1/3$$

to match the net flux in and maintain stable populations

$$j_n = j_1 - kS_1 = j_1 - k\left[(2j_1)/3k\right] = j_1/3$$

The stable populations for a steady flux define and entropy for each channel in terms of their fractional populations, i.e., probabilities

$$S = -k\{f_1 \ln f_1 + f_s \ln f_s\}$$

The steady flow through the system has reduced the entropy of the system relative to the equilibrium state when both fluxes are equal

$$S = -k[1/2 \ln 1/2 + 1/2 \ln 1/2] = k \ln 2$$

since the fractions are different.

17.3 General Linearized Driven Systems

Channels are good models for driven systems because they limit the number of possible transitions. For a three-site channel, the ion moves from site 1 to site 2 but not from site 1 to site 3. For a three-site channel with nearest neighbor site transitions and equal rate constants, the matrix is tridiagonal

$$K = \begin{pmatrix} -2k & k & 0 \\ k & -2k & k \\ 0 & k & -2k \end{pmatrix}$$

The general K $N \times N$ matrix might allow transitions between all states. However, the stationary state populations are still

$$|s >= -\mathbf{K}^{-1}|j >$$

the inverse for any matrix is found using eigenvalue-projection operator techniques where the projection operator Z_i is formed from the normalized eigenvectors

$$Z_i = |i >< I|$$

The inverse for K with eigenvalues is

$$\mathbf{K}^{-1} = \Sigma \lambda_i^{-1} Z_i$$

The tridiagonal matrix has eigenvalues and normalized eigenvectors

$$\left(-2 + \sqrt{2}\right) k \qquad (1/2)\left(1, \sqrt{2}, 1\right)$$

$$2 k \qquad (1/2)\,(2,0,-2)$$

$$\left(-2 - \sqrt{2}\right) k \qquad (1/2)\left(1, -\sqrt{2}, 1\right)$$

and inverse

$$\frac{1}{4k\left(--2+\sqrt{2}\right)} \begin{pmatrix} 1 \\ \sqrt{2} \\ 1 \end{pmatrix} \left(1 \ \sqrt{2} \ 1\right) + \frac{1}{-8} \begin{pmatrix} 2 \\ 0 \\ -2 \end{pmatrix} \left(2 \ 0 \ -2\right)$$

$$+ \frac{1}{4k\left(-2-\sqrt{2}\right)} \begin{pmatrix} 1 \\ \sqrt{2} \\ 1 \end{pmatrix} \left(1 \ -\sqrt{2} \ 1\right)$$

The row eigenvector of the inverse matrix forms a scalar product with the flux vector

$$|j >= \begin{pmatrix} j_1 \\ 0 \\ j_2 \end{pmatrix}$$

to give the three populations

$$\left(\frac{j_1 + j_2}{4k\left(2 - \sqrt{2}\right)}\right) \begin{pmatrix} 1 \\ \sqrt{2} \\ 1 \end{pmatrix} + \frac{j_1 - j_2}{8k} \begin{pmatrix} 2 \\ 0 \\ -2 \end{pmatrix} + \frac{j_1 + j_2}{4k\left(2 + \sqrt{2}\right)} \begin{pmatrix} 1 \\ -\sqrt{2} \\ 1 \end{pmatrix}$$

The first eigenvalue is smallest and the first term is largest. The stationary state distribution for systems with many states can often be approximated satisfactorily by selecting only the term in the smallest eigenvalue.

The stationary flow through the channels is calculated in four different ways that give the same net flux:

(1) the net flux from bath 1 to Site 1; $j_1 - k(S_1)$
(2) the net flux from site 1 to site 2; $k(S_1) - k(S_2)$
(3) the net flux from site 2 to site 3; $k(S_2) - k(S_3)$
(4) the net flux from site 3 to bath 2; $k(S_3) - j_2$

For any system, the total population is the sum of all state populations and is used to form the population fractions for the system entropy.

For a channel where the sites are different, the rate constants can differ. The rate constant matrix, while more complicated, can still be inverted and used to establish the site populations.

17.4 The Driven Wind-Tree Model

The wind-tree model is a four state linear kinetic system that uses transitions probabilities. The system evolves to an equilibrium where a wind particle has equal probabilities in all four directions. The wind-tree model is converted into a driven system when wind particles are added to the plane in one direction, e.g., north to south (direction 1), while an equal number is removed from the south to north direction (direction 3). The total wind particles are conserved but the stationary state distribution includes different populations.

The equilibrium wind-tree equations

$$df_1/dt = -2kf_1 + kf_2 + kf_4 = 0$$
$$df_2/dt = kf_1 - 2kf_2 + kf_3 = 0$$
$$df_3/dt = kf_2 - 2kf_3 + kf_4 = 0$$
$$df_4/dt = kf_1 + kf_3 - 2kf_4 = 0$$

are modified with input rate j (direction 1) and output rate $-j$ (direction 3)

$$df_1/dt = -2kf_1 + kf_2 + kf_3 + j = 0$$
$$df_2/dt = kf_1 - 2kf_2 + kf_3 = 0$$
$$df_3/dt = kf_2 - 2kf_3 + kf_4 - j = 0$$
$$df_4/dt = kf_3 + kf_1 - 2kf_4 = 0$$

or, at stationary state,

$$d/dt \begin{pmatrix} f_1 \\ f_2 \\ f_3 \\ f_4 \end{pmatrix} = \begin{pmatrix} -2k & k & 0 & k \\ k & -2k & k & 0 \\ 0 & k & -2k & k \\ k & 0 & k & -2k \end{pmatrix} \begin{pmatrix} f_1 \\ f_2 \\ f_3 \\ f_4 \end{pmatrix} + \begin{pmatrix} j \\ 0 \\ -j \\ 0 \end{pmatrix} = 0$$

$$d|f>/dt = \mathbf{K}|f> + |j> = 0$$
$$|f> = \mathbf{K}^{-1}|j>$$

is the sum of an equilibrium vector, doubly degenerate projection operator matrix with eigenvalue $2\,k$ and a projection operator matrix with eigenvalue $4\,k$

$$\mathbf{K}^{-1} = |c^{eq}> + \sum \frac{1}{\lambda_i}|i><i|$$

$$= |c> = \begin{pmatrix} .25 \\ .25 \\ .25 \\ .25 \end{pmatrix}$$

$$+ \frac{j}{4k} \begin{pmatrix} 1 & 0 & -1 & 0 \\ 0 & 1 & 0 & -1 \\ -1 & 0 & 1 & 0 \\ 0 & -1 & 0 & 1 \end{pmatrix} \begin{pmatrix} 1 \\ 0 \\ -1 \\ 0 \end{pmatrix} + \frac{j}{8k} \begin{pmatrix} 1 & -1 & 1 & -1 \\ -1 & 1 & -1 & 1 \\ 1 & -1 & 1 & -1 \\ -1 & 1 & -1 & 1 \end{pmatrix} \begin{pmatrix} 1 \\ 0 \\ -1 \\ 0 \end{pmatrix}$$

$$|f> = \begin{pmatrix} .25 \\ .25 \\ .25 \\ .25 \end{pmatrix} + \frac{j}{2k} \begin{pmatrix} 1 \\ 0 \\ -1 \\ 0 \end{pmatrix}$$

The fraction of wind particles in the north to south direction (1) increases as the south to north fraction decreases. Directions 2 and 4 have unchanged populations

$$f_1 = 0.25 + b$$
$$f_2 = 0.25$$
$$f_3 = 0.25 - b$$
$$f_4 = 0.25$$

The stationary state populations are stable as long as the flow is maintained. The wind particles do not pass sequentially through states as they would for the deterministic model. However, the system can be modified to produce a cyclic progression through the states by changing the shape of the trees so they are wedge shaped.

Wind particles change from 1 to 2, 2 to 3, etc., but not 1 to 4, 4 to 3, etc. The K matrix

$$\mathbf{K} = k \begin{pmatrix} -1 & 0 & 0 & 1 \\ 1 & -1 & 0 & 0 \\ 0 & 1 & -1 & 0 \\ 0 & 0 & 1 & -1 \end{pmatrix}$$

has equilibrium fractions (0.25, 0.25, 0.25, 0.25). However, the non-zero eigenvalues for the matrix contain the imaginary i

$$\lambda = 0, -1 + I, -2, -1 + i$$

The eigenvalues with imaginary terms will produce a decay with periodic behavior

$$\exp(-k(-1+i)t)$$

The real terms in the eigenvalues insure that this system will decay to the fully random distribution in time. The imaginary term indicates that it will spiral cyclically through the states as it evolves to this equilibrium. This oscillatory behavior is only observed if the input and output directions are selected to give a non-zero scalar product. For the eigenvalue -1+i, an input at 1 with output at 3 produces this cycle to equilibrium

$$< -1 + i|j> = \left(1 \ -i \ -1 \ i\right) \begin{pmatrix} j \\ 0 \\ -j \\ 0 \end{pmatrix} = 2j$$

The modified wind tree cannot sustain cyclic oscillations when driven by the input and output of particles. However, it can still be maintained in a non-equilibrium state with the proper input and outputs. The eigenvalue -2 with eigenvector $(1, -1, 1, -1)$ which is part of the inverse matrix expansion will maintain a steady non-equilibrium distribution for the input flux $(j,0,0, -j)$.

17.5 The Entropy Decrease in Driven Systems

Directional flow asserts an order on a system to produce a decrease in the system entropy. A system with A and B isomers of equal energy (the dog-flea model) has equal populations of A and B at equilibrium. Since

$$A \ \underset{k}{\overset{k}{\rightleftharpoons}} \ B$$

$$dA/dt = 0 = -k(A) + k(N - A)$$

and

$$f_A = A/N = 0.5 \quad f_B = B/N = 0.5$$

The system is now pumped to a stationary state by adding A at a rate r, while removing B at the same rate. The stationary state concentrations are

$$|dA/dt = 0 = r - kA + kB$$
$$dB/dt = 0 = -r + kA - kB$$

$$0 = \begin{pmatrix} r \\ -r \end{pmatrix} - k \begin{pmatrix} 1 & -1 \\ -1 & 1 \end{pmatrix} \begin{pmatrix} A \\ B \end{pmatrix}$$

The populations are the sum of an equilibrium vector and a perturbation vector in r/k

$$\begin{pmatrix} f_A \\ f_B \end{pmatrix} = \begin{pmatrix} 0.5 \\ 0.5 \end{pmatrix} + \frac{1}{2k} \begin{pmatrix} 1 \\ -1 \end{pmatrix} (1 -1) \begin{pmatrix} r \\ -r \end{pmatrix}$$

$$f_A = 0.5 + r/k$$
$$f_B = 0.5 - r/k$$

The parameter r/k, which must be smaller than 0.5, is a measure of the distance from equilibrium.

The entropy at stationary state is determined as

$$-S/k = f_A \ln f_A + f_B \ln f_B = (0.5 + r/k) \ln (0.5 + r/k) + (0.5 - r/k) \ln (0.5 - r/k)$$
$$= (0.5 + r/k) \left[\ln (0.5) + \ln (1 + 2r/k) \right]$$
$$+ (0.5 - r/k) \left[\ln (0.5) + \ln (1 - 2r/k) \right]$$

Using $\ln(1+x) = x$, the entropy is

$$S/k = \ln (2) - 2(r/k)^2$$

The linear terms in r/k cancel so that the entropy decreases from its equilibrium maximum as the square of r/k.

The entropy reduction for a driven wind-tree model where the wind is added in the 1 direction and removed at 3 has a stationary state correction that uses only the $2k$ eigenvalue

$$\begin{pmatrix} f_1 \\ f_2 \\ f_3 \\ f \end{pmatrix} = \begin{pmatrix} 0.25 \\ 0.25 \\ 0.25 \\ 0.25 \end{pmatrix} + \frac{r}{2k} \begin{pmatrix} 1 \\ 0 \\ -1 \\ 0 \end{pmatrix} (1 \ 0 \ -1 \ 0) \begin{pmatrix} r \\ 0 \\ -r \\ 0 \end{pmatrix} = \begin{pmatrix} 0.25 + r/k \\ 0.25 \\ 0.25 - r/k \\ 0.25 \end{pmatrix}$$

With opposite deviations in only two states, the entropy decrease for this four-state system is

$$S/k = \ln (4) - (2r/k)^2$$

The situation is more complicated for an open system like a membrane channel that reaches a dynamic equilibrium only if the inputs from both baths are j and the ion can return to the baths

$$dA/dt = r - 2kA + kB$$
$$dB/dt = r + kA - 2kB$$

$$0 = -k \begin{pmatrix} 2 & -1 \\ -1 & 2 \end{pmatrix} \begin{pmatrix} A \\ B \end{pmatrix} + \begin{pmatrix} j \\ j \end{pmatrix}$$

$$\begin{pmatrix} A \\ B \end{pmatrix} = K^{-1} |j> = \frac{1}{3k} \begin{pmatrix} 2 & 1 \\ 1 & 2 \end{pmatrix} \begin{pmatrix} j \\ j \end{pmatrix} = \frac{r}{3k} \begin{pmatrix} 3 \\ 3 \end{pmatrix} = \begin{pmatrix} r/k \\ r/k \end{pmatrix}$$

For this open model, the populations of both sites are determined by the input flux j; the populations are equal when the fluxes from both sides are equal.

With a flux only to site 1, the populations are

$$\begin{pmatrix} A \\ B \end{pmatrix} = \frac{1}{3k} \begin{pmatrix} 2 & 1 \\ 1 & 2 \end{pmatrix} \begin{pmatrix} j \\ 0 \end{pmatrix} = \begin{pmatrix} 2j/3k \\ j/3k \end{pmatrix}$$

The population of site 1 is now twice that of site 2.

The entropy is determined by breaking the population into a common concentration and fractional populations at each site. For this case, the concentration is j/k and the fractional populations are 2/3 and 1/3 respectively. The entropy is

$$(j/k)\left[\frac{2}{3} \ln\left(\frac{2}{3}\right) + \frac{1}{3}\ln\left(\frac{1}{3}\right)\right]$$

17.6 Lasers

Flow systems with linear kinetics are driven from equilibrium with a redistribution of states. The basic kinetic mechanisms remain at all drive levels. The mechanism for lasing involves a transition from linear first-order to a non-linear second-order mechanism. As the input flux of excitation photons increases, the emission mechanism changes from spontaneous emission by single molecules to stimulated emission where the emission photon from one molecule stimulates the emission of a photon from a second molecule.

A molecule absorbs photons for a transition from its ground state S_o to an excited state S^*. The rate depends on the S_o molecules and photons expressed as an intensity I

$$k_e S_o I$$

Normally, $S_o \gg I$ and the rate is pseudo-first order in I

$$(k_e S_o)I = k'I$$

This rate is similar to the input flux j used to produce populated membrane channels.

Although the excited S^* can react via several pathways, for this example, S^* is depopulated only by fluorescence, the first-order emission of a photon to restore the ground state S_o with rate constant $k_f S^*$.

The stationary state S^* population is

$$0 = k_e S_o I - k_f S^*$$
$$S^* = k_e S_o I / k_f$$

The rate of emission (fluorescence) equals the rate from S^* to S_o

$$I_f = k_r S^* = k_e S_o I$$

Since fluorescence is the sole depopulation path, all added photons reappear as fluorescent photons.

The spontaneous behavior of emission is apparent if the S^* are all created with a short pulse of photons (a delta function) at $t=0$. The initial population $S^*(0)$ decays as

$$S^*(t) = S^*(0)\exp(-k_f t)$$

Einstein used the concept of microscopic reversibility to postulate stimulated emission. An excitation is second order in S_0 and I. One reverse path from S^* to S_0 must also be second order in both S^* and I. The rate constants for excitation and stimulated emission are equal because of microscopic reversibility. The non-linear stimulated emission rate

$$k_{se}S^*I = kS^*I$$

is included in the stationary state rate equation for S^*

$$0 = kS_0I - (kI + k_f)S^*$$
$$S^* = kS_0I/(kI + k_f)$$

The rates of spontaneous and stimulated emission, respectively, are

$$r_s = k_f S^* = k_f k S_0 I/(kI + k_f)$$
$$r_{se} = kIS^* = (kI)^2 S_0/(kI + k_f)$$

An increase in I favors stimulated emission. Lasing occurs with a population inversion where $S^* > S_0$. However,

$$S^*/S_0 = kI/(kI + k_f) < 1$$

for finite k_f. Most lasers operate by transferring the energy to S^* from A molecule with higher energy. A helium–neon laser uses the excited state of He to populate an excited state of neon and a neon population inversion.

Stimulated emission when $S^* > S_0$ produces cooperative behavior. S^* are created along the longitudinal axis of a laser. A photon emitted along this axis passes other S^* to stimulate their emission with the same direction and phase. The two photons continue along the axis to stimulate additional S^*. Mirrors extend the photon path along the axis to stimulate additional S^* to emit. For an ideal laser, the photons emitted move in the same direction with equal phase. The large drive photon population necessary to create the inversion produces order in the system. The population inversion is the threshold between spontaneous and stimulated emission.

17.7 The Nerve Impulse

Sodium and potassium channels in a nerve cell combine to produce a simple one shot oscillator. A resting nerve axon has channels for Na^+ that are closed so that the +110 mV voltage that these ions can generate as an electrical potential is absent.

The potassium channels in the same axon are sufficiently open to create an internal −60 mV potential from the K$^+$ gradient (c_K(inside) > c_K(outside)). This negative potential is observed for the unstimulated axon. The "equilibrium" potential is stable even if small currents are injected. The potential forced to −50 mV by a current injection pulse decays back to −60 mV when the current pulse terminates.

The situation changes dramatically if the injection current drives the membrane potential beyond a threshold. Instead of decaying exponentially back to −60 mV when the pulse terminates, the potential continues to move away from −60 mV. The potential increases from −60 to +50 mV and then begins to relax back to negative potentials in a period of about 2 ms (Fig. 17.4).

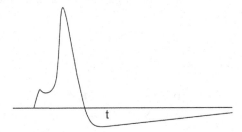

Fig. 17.4 The nerve action potential

The system moves away from equilibrium when the injection current produces a voltage change sufficient to induce opening of some sodium channels; the sodium potential (+110) then begins to equilibrate raising the potential still more. This "autocatalytic" change in the number of open sodium channels produces a rapid rise with a concomitant addition of the full sodium ion potential of +110 mV. The observed peak at voltage 50 mV (−60 mV + 110mV) produces two effects: (1) a second mechanistic step closes the sodium channels cutting off their 110 mV while (2) a greater fraction of potassium channels also open so the potential actually drops to −72 mV – the electrochemical potential observed with open potassium channels. The system then re-equilibrates to −60 mV in preparation for the next stimulus.

The initial potential of −60 mV is produced by the permeabilities or mobilities of K$^+$ and Na$^+$ in their respective ion channels via the Goldman Hodgkin Katz equation (the sum of the constant field equations for Na$^+$ and K$^+$) with the electrical mobility for sodium and potassium ion in their respective channels

$$\psi = -[kT/e]\ln\left[\frac{u_K K_2 + u_{Na} Na_2}{u_K K_1 + u_{Na} Na_1}\right]$$

For the initial equilibrium, $u_K > u_{Na}$, the potassium potential

$$V = -\frac{kT}{e}\ln\frac{c_K(inside)}{c_K(outside)}$$

dominates. The spike potential from Na$^+$ dominates when $u_{Na} > u_K$ because the sodium channels are open.

The open-closed kinetic sequence built into the sodium channels produces a one shot oscillation. The axon leaves it equilibrium potential of -50 mV to reach a new transient potential of $+50$ mV before returning to its initial voltage of -60 mV. The sodium–potassium channel combination produces an oscillating system driven by the concentration gradients of Na^+ and K^+ across the membrane. The contents of the axon can be extruded and replaced by an electrolytic solution of the proper concentrations and the axon functions perfectly. The oscillatory behavior resides entirely in the kinetic behavior of the sodium and potassium channels.

17.8 The Prey–Predator Model

A train of nerve impulses is an oscillating reaction where the populations of open sodium and potassium channels change in both time and phase. The transition between two equilibria for the oscillations is an autocatalytic process where the presence of open sodium channels triggers the formation of additional sodium channels. The prey–predator model is an example of a oscillating system where the non-linear behavior produces non-equilibrium changes in population

Prey–predator kinetics postulates first-order growth for the prey, e.g., rabbits (R) and a second-order rate of loss proportional to the populations of predator(foxes F) and prey (rabbits R). The R rate equation

$$\frac{dR}{dt} = k_r R - k_{-r} R F$$

is coupled to a rate of change of foxes whose increase is proportional to both the existing foxes and rabbits, i.e., second order or non-linear while the rate of loss for foxes (without their own predators) is proportional to the fox population

$$\frac{df}{dt} = k_f FR - k_{-1} F$$

The system has a stable equilibrium when both rates are 0

$$0 = k_r R - k_{-r} R F$$

and

$$0 = k_f F R - k_{-f} F$$

$$F = \frac{k_r}{k_{-r}} \qquad R = k_{-f}/k_f$$

$$= k_f F^e R^e - K_{-f} F^e + K_{ff} R + k_{fr} F - k_{-ff}$$

The system does not always reach these equilibrium populations. A small change in these equilibrium populations can trigger oscillatory behavior around equilibrium. If the equilibrium population of rabbits increases by r,

$$R = R^e + r$$

while the fox population changes by f,

$$F = F^e + f$$

the rate equation for r is

$$\frac{d(R^e + r)}{dt} = \frac{dr}{dt} = k_r(R + r) - k_{-r}(R + r)(F + f)$$

while the rate of change for f is

$$\frac{d(F^e + f)}{dt} = \frac{df}{dt} = k_f(F^e + f)(R^e + r) - k_{-f}(F^e + f)$$

The equilibrium rates for in R and F are zero. For small perturbations from equilibrium, rf is small and equated to zero. The final rate equations in the perturbed populations r and f

$$\frac{dr}{dt} = k_r r - k_{-r}\left(r\frac{k_r}{k_{-r}}\right) - k_{-r}\left(f\frac{k_{-f}}{k_f}\right) = -\frac{k_{-f}k_{-r}}{k_f}f$$

$$\frac{df}{dt} = k_f^e(RF^e + rF^e + fR^e + rf) - k_{-f}F - k_{-f}f$$

$$\frac{df}{dt} = k_f r\frac{k_r}{k_{-r}} + k_f f\frac{k_{-f}}{k_f} - k_{-f}f = \frac{k_f k_r}{k_{-r}}r$$

The two equations are coupled. dr/dt depends on f while df/dt depends on r.

The two equations combine to reduce to a single second-order differential equation

$$\frac{d^2 r}{dt^2} = -\frac{k_{-r}k_{-f}}{k_f}\frac{df}{dt} = -\frac{k_{-r}k_{-f}}{k_f}\frac{k_f k_r}{k_{-r}}r = -k_{-f}k_r r$$

with a sine or cosine solution. The population of excess rabbits r oscillates in time as

$$r = \sin \omega t$$

with frequency of oscillation

$$\omega = \sqrt{k_{-f}\, k_r}$$

to satisfy the differential equation.

The excess fox population oscillates with the same frequency. However, it lags the rabbit population because the population of rabbits must increase before the fox population begins to increase.

As the rabbit population increases the fox population changes slowly. When the rabbit population increases, the fox population begins to rise rapidly. The excess foxes then decrease the rabbit population so that the fox population again decreases.

This cycle can continue indefinitely. The system is not at equilibrium; it exists as an oscillating stationary state about equilibrium.

More sophisticated prey–predator models do not assume homogeneous populations of prey and predator. A population of rabbits might develop as a fluctuation in some sub-area and their population spreads from this region. A color map of rabbits (red) and foxes (blue) shows regions of blue appearing and growing while, with a phase delay, red impinges and "grows" into the blue areas. The behavior is similar to that observed for oscillating chemical reactions in a flask.

17.9 Oscillating Reactions

An autocatalytic reaction needs an autocatalytic step where a reactant B produced at some steady rate exceeds a threshold to produce a rapid increase in the product. When the available B is exhausted in the reaction, other kinetics deplete this molecule to return the initial conditions so a new cycle can begin.

The basic mechanism is illustrated by a model with the key steps:

(1) The reaction is "fueled" by the steady formation I of one of the reactants, $S \rightarrow A$
(2) The catalytic material (B) is created from A by an uncatalyzed slow, first-order reaction step

$$k_1[A]$$

(3) An autocatalytic step of order n in B gives a strong rate increase when B reaches a critical concentration. For $n = 2$

$$A + 2B \rightarrow 3B$$
$$R = k_c[A][B]^2$$

Once this autocatalytic step begins, B increases to a large concentration depleting the reactant A that sustains the reaction.
(4) No additional B is produced by autocatalysis and its concentration decreases via a slower irreversible reaction to product $-k_2[B]$

This chemical flow system is driven by the chemical reactions that carry the starting reactant A to a final product. The energy for the oscillations is tapped from the energy difference of the reactants and products.

A does continue to form and can build to trigger a second autocatalytic cycle; the cycles continue until production of A stops.

The rate equation for A includes the steady input of A from the source S and the loss of A by both an irreversible first-order reaction and the irreversible autocatalytic reaction,

$$\frac{d[A]}{dt} = I - k_1[A] - k_x[A][B]^2$$
$$A^{ss} = \frac{I}{k_x[B]^2 + k_1}$$

The creation of B occurs via both the first-order reaction of A and the third-order autocatalytic reaction. The reaction rate is decreased by the loss of B to product C ($k_2[B]$),

$$\frac{d[B]}{dt} = k_c[A][B]^2 + k[A] - k_2[B]$$

Steady-state concentrations of A and B are determined by setting the rates to zero and adding

$$0 = I - k_c[A][B]^2 - k_1[A] + (k_c[A][B]^2 + k_1[A] - k_2[B]) = I - k_2[B]$$

The steady-state concentrations of B and A, respectively, are

$$B^{ss} = \frac{I}{k_2}$$

$$I - \left(k_c[B]^2 + k_1[A]\right) = 0$$

The concentrations are perturbed from their steady-state values by concentrations a and b for the A and B molecules, respectively,

$$[A(t)] = A^{ss} + a$$

$$[B(t)] = B^{ss} + b$$

The rate equation for a is

$$\frac{d(A^{ss} + a)}{dt} = \frac{da}{dt} = I - k_c(A^{ss} + a)(B^{ss} + b)^2 - k_1\left(A^{ss} + a\right)$$

The corresponding equation for b is

$$\frac{db}{dt} = I - k_c A^{ss}(B^{ss})^2 - k_1 A^{ss}\# - k_c a B^{ss2} - k_c 2A^{ss}B^{ss}b - k_1 a + \dots$$

Combining these two equations for one equation in a or b does not give a sinusoidal like the prey–predator model but does oscillate in time.

17.10 Resonance and Stochastic Resonance

Driven systems acquire order reflected in a decrease in the system entropy when subjected to a constant directed input. The order depends on the time independent ratio of flux to rate constant j/k. Order can also develop from a time dependent input with period consistent with the rate constants of the system. The system gives maximal response to the input when the frequency of the input matches characteristic frequencies of motions within the system. Although the frequencies involve all the eigenvalues of the system, the smallest eigenvalue often dominates. A system with a single characteristic frequency is developed as an example. The system is resonant when the oscillation period matches a period of the system.

A light beam is intensity modulated at a frequency $\omega = 2\pi/T$ where T is comparable to the dominant system time constant τ. If the modulated photon beam produces a modulated concentration of excited S^*, the products of S^* are also modulated at this frequency although phase delayed relative to the input. The kinetic scheme

$$I(t) + S \xrightarrow{k_e} S^*(t) \xrightarrow{k_f + k'} P(t) + I_f(t)$$

where k_f and k' are first-order constants for fluorescent and non-fluorescent depopulations, respectively. If $k = k_f + k'$ and $I(t) = I \sin(\omega t)$, the rate is

$$dS^*/dt = j = k_e Ij \sin(\omega t) - kS^*$$

With integrating factor $\exp(kt)$

$$dS^*/dt \exp(kt) + kS^* \exp(kt) = k_e I \sin(\omega t) \exp(kt)$$

$$\int_{S^*=0,t=0}^{S^*(t),t} d[S^* \exp(kt)] = k_e I \int_0^t \sin(\omega t) \exp(kt) dt$$

$$S^*(t) \exp(kt) - 0 = k_e I \exp(kt)[k \sin(\omega t) - \omega \cos(\omega t)]/[k^2 + \omega^2]_0^t$$

$$S^*(t) = k_e I[k \sin(\omega t) - \omega \cos(\omega t)]/(k^2 + \omega^2)$$

$$+ \omega \exp(-kt)/(k^2 + \omega^2)$$

The exponential term decays to leave a stationary state in sine and cosine functions. However, the definitions

$$\cos(\varphi) = k/\sqrt{(k^2 + \omega^2)} \qquad \sin(\varphi) = \omega/\sqrt{k^2 + \omega^2}$$

are substituted to give

$$S^* = k_e I/[\cos(\varphi) \sin(\omega t) - \sin(\varphi) \cos(\omega t)]$$

$$= \frac{k_e I}{\sqrt{k^2 + \omega^2}} \sin(\omega t - \varphi)$$

and

$$\tan(\varphi) = \sin(\varphi)/\cos(\varphi) = \omega/k = \omega\tau$$

The fluorescence intensity is also modulated

$$I_f = k_f S^*(t) = k_e I k_f \sin(\omega t - \varphi)/\sqrt{k^2 + \omega^2}$$

The output intensity falls with increasing frequency while the system period T equals the driving period at 45°.

The maximal signal at the resonant frequency suggests that all molecules are emitting in synchrony. If they are excited by a pulse of light, the decay is exponential indicating that each molecule emits independently with no "memory." The resonant frequency imparts a common time bias to all the emitting molecules.

Stochastic resonance is observed in non-linear systems driven by both a periodic signal at any frequency and a range of noise frequencies. Both components below a threshold necessary for a transition to a second stable state but transitions between states occur. For example, a strong periodic signal can promote transitions between the wells of a two-well potential with non-linear terms

$$V(x) = -kx^2 + ax^4$$

The same periodic signal with amplitude and attendant noise below threshold promotes the same transitions. This is stochastic resonance.

17.11 Synchronization

The systems driven by an external periodic input fall into synchronization with that frequency to give a constant phase shift for the output. Systems can also spontaneously reach a common frequency and phase with no controlling external input, a phenomenon called synchronization. The classic example of synchronization is that of two separate clock pendula mounted on a common bar. The two pendula reach a common period with a constant phase of 180; both pendula move inward at the same time. . Their natural periods differ slightly because its almost impossible to produce pendulums an exact frequency match. However, the periods and phases of both clocks shift until the clocks have exactly the same period and are exactly 180° out of phase.

The stimulated emission of a laser is an non-linear process depending on both the number of photons and the excited states. A photon released from a molecule in the laser cavity induces emission from other excited molecules. Although such molecules might be released with a range of wavelengths, the interaction between the photon and the excited molecule produces a second photon of the same phase and wavelength.

A single pendulum need not force its frequency on the second. Both pendula have frequencies within some capture range and both change to reach a common frequency. A frequency difference also appears as a phase difference. A signal at higher frequency moves through its phases more rapidly to produce a phase shift relative to the second, lower frequency signal. Synchronization is mapped using a phase plot. Sinusoidal signals have a phase changing as ωt, where ω is the angular frequency. The phase change is plotted as a point on a circle for the periodic system. One complete cycle moves the point 360° around the circle. For two synchronizing units, a second point on the circle describes the second unit. If the second frequency is higher, its phase point moves more rapidly around the circle. As the two signals come into synchronization, they move around the circle at the same rate although their phase may differ. For example, two synchronized pendula with periodic motion have a plot with two points that move about the circle separated by 180°. Before synchronization, one phase moves ahead or behind the second. Their separation becomes constant only when the two systems are synchronized.

Problem Solutions

Chapter 1

1.1 Solve for P, insert in PdV integral

$$w = -\int \frac{RT}{V-b} - \frac{a}{V^2} dV$$

$$w = -\int \left(\frac{RT}{V-b} - \frac{a}{TV^2} \right)$$

$$w = -\int \left(\frac{RT}{V-b} \exp\left(-a'/RTV\right) \right) dV$$

1.2 a. $w = -nRT \ln(V_f/V_i) = -PV \ln(V_f/V_i) = -100 \ln(100/1) = -4600$ Latm
b. $w = -1(100-1) = -99$ Latm
c. $w = -50(2-1)-1(100-2) = -138$ Latm

1.3 $\Delta E = q + w = -12.1 - 8 = -20.1$ Latm

$$w = -P_{ext}\Delta V$$

$$-8 = -2(\Delta V) \quad \Delta V = +4$$

$$\Delta T = \Delta E/(3(R)/2) = -2000J/12.5 = -160$$

1.4 a. $w = -RT \ln \left(\frac{V_2 - b}{V_1 - b} \right)$
b. $w = -P_{ext}(V_2 - V_1)$

M.E. Starzak, *Energy and Entropy*, DOI 10.1007/978-0-387-77823-5,
© Springer Science+Business Media, LLC 2010

1.5 $\quad w = -RT \ln (V_2/V_1) = -RT \ln \left(\dfrac{RT/P_2}{RT/P_1}\right) = +RT \ln (P_2/P_1)$

1.6 $\quad C_p = 5R/2 = 20 \quad \Delta T = \Delta H/C_p = 2500/20 = 125°$
$\quad\quad C_v = C_p - R = 12.5$

$$\Delta E = C_v \Delta T = 12.5(125) = 1562.5 \text{ J}$$

1.7 $\quad \Delta E = -2500 = 0 + w \quad\quad w = -2500$

$$\Delta T = -2500/12.5 = -200 \quad\quad \Delta H = C_p \Delta T = 20(-200) = -4000$$

1.8 $\quad F = \tau$ for reversible work

$$w_{rev} = \int \tau \, dL = \int\limits_{x_o}^{2x_o} k(x - x_o) dx = (k/2)x_o^2$$

$$\Delta E = w \text{ since } q = 0$$

Chapter 2

2.1 a. Cool $H_2 + Br_2$ from 400 to 300++
     b. Change $H_2 + Br_2$ to $2HBr$ at 300
     c. Heat $2HBr$ to 400

$\quad\quad C_p = 5R/2$ for H_2, I_2 and HI
$\quad\quad 2C_p(\text{diatomic})(300-400) - 38 \text{ kJ} + 2C_p(\text{diatomic})(400-300) = -28 \text{ kJ}$

2.2 $\quad w = -P_{ext} (V_f - V_i) = -1(2-1) = -1 \text{ Latm} = -100 \text{ J}$

$$\Delta E = C_v \Delta T + (-0.1 \text{ Latm}) = 5R/2(0) - 0.1 \text{ Latm} = -0.1 \text{ Latm}$$

$$q = \Delta E - w = -0.1 \text{ Latm} - (-1 \text{ Latm}) = 0.9 \text{ Latm}$$

2.3 $\quad q = 5000 \text{ J for step 1} \quad$ since $q = 0$ for the adiabatic step 2

$$\Delta E = 0 \text{ for step 1} \quad \text{so } w = -q = -5000$$

$$q = 0 \text{ for step2} \quad\quad \Delta E = -500 \text{ J so } \Delta E = w = -500 \text{ J for step 2}$$

2.4 $C_v(T_2-T_1)+a(V_2-V_1)$

2.5 $H(A) = +1000$, i.e. 1000 over the reference for B as zero

2.6 $12.5(300-350) + 20(11-1) = \Delta E = 625 + 200 = 825$, $w = +0.2(11-1) = 2$
Latm ≈ 200 J

$$q = \Delta E - w = 825 - 200 = 625J$$

$$\Delta H = (12.5 + R)\Delta T = 1000$$

2.7 $\Delta H = 1000 = 20\Delta T \qquad \Delta = +50$

$$\Delta E = 12.5(50) = 625$$

Chapter 3

3.1 a. Since ideal, $\left(\dfrac{\partial H}{\partial P}\right)_T = 0 \qquad dT/dP = 0/C_p = 0$

b. $\Delta E = 0$ since isothermal and ideal

3.2 $\Delta H = 4000 = (20)\Delta T \qquad \Delta T = 4000/200 = 200$

$$\Delta E = 12.5(200) = 2500$$

3.3 $w = -907 \qquad q = -347 \qquad \Delta E = q + w = -907 - 347 = -1254$

$$\Delta T = -1254/12.5 = 100.3$$

3.4 a. Since $q = 0$, $\Delta E = 0 + w = -100$

b. $q = 0 \qquad \Delta T = -100/12.5 = -8 \qquad \Delta H = 20(-8) = -160$

c. $T_f = 300 + (-8) = 292$

3.5 a. $\left(\dfrac{\partial H}{\partial T}\right)_P = \left(\dfrac{\partial E}{\partial T}\right)_P + P\left(\dfrac{\partial V}{\partial T}\right)_P$

$$\left(\frac{\partial E}{\partial T}\right)_P = \left(\frac{\partial E}{\partial T}\right)_V \left(\frac{\partial T}{\partial T}\right)_P + \left(\frac{\partial E}{\partial V}\right)_T \left(\frac{\partial V}{\partial T}\right)_P$$

$$\left(\frac{\partial H}{\partial T}\right)_P = \left(\frac{\partial E}{\partial T}\right)_V + \left[\left(\frac{\partial E}{\partial V}\right)_T + P\right]\left(\frac{\partial V}{\partial T}\right)_P$$

b. $\left(\dfrac{\partial E}{\partial T}\right)_V = \left(\dfrac{\partial H}{\partial T}\right)_V - V\left(\dfrac{\partial P}{\partial T}\right)_V$

$\left(\dfrac{\partial H}{\partial T}\right)_V = \left(\dfrac{\partial H}{\partial T}\right)_P\left(\dfrac{\partial T}{\partial T}\right)_V + \left(\dfrac{\partial H}{\partial P}\right)_T\left(\dfrac{\partial P}{\partial T}\right)_V$

$\left(\dfrac{\partial E}{\partial T}\right)_V = \left(\dfrac{\partial H}{\partial T}\right)_P - \left[-\left(\dfrac{\partial H}{\partial P}\right)_T + V\right]\left(\dfrac{\partial P}{\partial T}\right)_V$

3.6 $\alpha = \dfrac{1}{V}\partial(RT/P)/\partial T = R/PV = 1/T$

$\beta = \dfrac{1}{V}\partial(RT/P)/\partial P = -RT/VP^2 = -P/P^2 = -1/P$

3.7 $C_V dT = -(RT/V)dV$ $C_V dT/T = -RdV/V$

$C_V \ln(T_f/300) = -R \ln(22.4/44.8)$ $1.5 \ln(T_f/300) = +\ln(2)$

$\mathrm{Ln}(T_f/300) = +0.46$ $T_f = 300\ \exp(0.46) = 475\mathrm{K}$

3.8 $D = H + RT$

a. Made of state functions

b. $H = D - RT$ $\left(\dfrac{\partial H}{\partial T}\right)_P = \left(\dfrac{\partial D}{\partial T}\right)_P - R$

c. $dD = 0 = \left(\dfrac{\partial D}{\partial T}\right)_P dT + \left(\dfrac{\partial D}{\partial P}\right)_T dP$

$(dT/dP)D = \left(\dfrac{\partial D}{\partial P}_T\right)\Big/\left(\dfrac{\partial D}{\partial T}\right)_P$

Chapter 4

4.1 S depends on volume change

$\Delta S = R \ln(V_f/V_i) = R \ln(5/2)$

4.2 $\Delta S = C_V \ln(300/350) + R \ln(11/1)$

$$w = -0.2(11-1) = -2\,\text{Latm}$$

$$q = \Delta E - w = 0 + 0.2(11-1) = +2\,\text{Latm} = +200\,\text{J}$$

$$\Delta S_{surr} = -200/300 = 2/3$$

4.3 $\Delta E = 0 + w = 12.5(300-350) = -625$ J
$\Delta S = 0$ for reversible adiabatic expansion

4.4   Reversible isothermal $\Delta S = R\,\ln(2)$; surroundings $\Delta S = -R\,\ln(2)$
Adiabatic reversible $\Delta S_{sys} = 0$ surroundings $\Delta S = 0$

4.5 $q_c/q_h = T_c/T_h = 270/300 = 0.9 = q_c/1200$ $\qquad q_c = 1080$; ideal work 120
50% does 240 J work $\qquad$ total heat added $1080 + 240 = 1320$ J

4.6   a. $\Delta S = 25\,\ln[300/400] = -7.1$

b. $q = 25(300-400) = -2500$ from block $\qquad$ 2500 to surroundings

c. $\Delta S_{univ} = 8.3 - 7.1 = +1.2$

4.7 $\Delta S = R\,\ln[(V_2-b)/(V_1-b)] + C_v\,\ln(T_2/T_1)$

4.8   a. $\Delta S_{sys} = 30\,\ln(400/200) - R\,\ln(0.25/1) =$ entropy of isothermal since adiabatic step $\Delta S = 0$

b. $\Delta S_{sur} = -\Delta S_{sys}$ since all is reversible

4.9 $\Delta S = -R\,\ln(P_2/P_1)$ for constant $T = -R\,\ln(0.25/1)$

Chapter 5

5.1   a. 0

b. 0 since Cl>>F only one orientation possible

5.2 $\Delta S = -R[0.1\,\ln(0.1) + 0.3\,\ln(0.3) + 0.6\,\ln(0.6)]$

5.3 $\Delta S_{mix}\,(400) = -5R[0.4\,\ln(0.4) + 0.6\,\ln(0.6)]$

Mixture to 300 $\qquad \Delta S = 5C_p\,\ln(300/400)$
5.4   Two entropies mixing and residual
$R\,\ln(2) - R[0.25\,\ln(0.25) + 0.75\,\ln(0.75)]$

5.5   a. $k\,\ln(1) = 0$

b. $k\,\ln(100)$

c. $k\,\ln(100)^5 = 5\,\ln(100)$
5.6   a. $S_{mix} = -4R[0.25\,\ln(0.25) + 0.75\,\ln(0.75)]$

b. $S_{mix} + C_p(A)\,\ln(400/200) + 3C_p(B)\,\ln(400/200)$
c. $1/4$

Chapter 6

6.1 $G = E + PV - TS \; dG = dE + PdV + VdP - TdS - SdT = -SdT + VdP$

6.2 $\Delta G = (H-TS)_f - (H-TS)_I = H_f - H_I - (TS)_f + (TS)_i$

$= 2000 - 1500 - (400 \times 3) + (300 \times 2) = -100$

6.3 $dG' = -SdT + VdP - Ad\gamma + qdV$

6.4 $\Delta H_{vap} = 342 \times 85 = 29{,}070$

$Ln(p/1) = -29{,}070/R[1/233 - 1/242]$

Chapter 7

7.1 $E/V = T(dP/dT) - P \quad 3P = T(dP/dT) - P$

$4P = T(dP/dT) \quad dP/P = 4dT/T \quad \ln P = 4\ln T = \ln T^4 \quad P\alpha T^4$

7.2 $dE = TdS + \tau dL$

a. $\left(\dfrac{\partial E}{\partial L}\right)_S = \tau$

b. $dG' = -SdT + \tau dL$

c. $\left(\dfrac{\partial S}{\partial L}\right)_T = -\left(\dfrac{\partial \tau}{\partial T}\right)_L \qquad \Delta S = -\int_{L1}^{L2}\left(\dfrac{\partial \tau}{\partial T}\right)dL$

7.3 $dG = -SdT + \tau dL \qquad \left(\dfrac{\partial S}{\partial L}\right)_T = -\left(\dfrac{\partial \tau}{\partial T}\right)_L$

$\left(\dfrac{\partial E}{\partial L}\right)_T = T\left(\dfrac{\partial S}{\partial L}\right)_T + \tau = -T\left(\dfrac{\partial \tau}{\partial T}\right)_L + \tau$

7.4 $\left(\dfrac{\partial E}{\partial V}\right)_T = T\left(\dfrac{\partial P}{\partial T}\right)V - P$

$\left(\dfrac{\partial P}{\partial T}\right)_V = d[RT/V - B]dT = \dfrac{RT}{(V-B)^2}\left(\dfrac{\partial B}{\partial T}\right) + R/V - B$

$\left(\dfrac{\partial E}{\partial V}\right)_T = \dfrac{RT^2}{(V-B)^2}\left(\dfrac{\partial B}{\partial T}\right) + \dfrac{R}{V-B} - \dfrac{R}{V-B} = \dfrac{RT^2}{(V-B)^2}\left(\dfrac{\partial B}{\partial T}\right)$

$\left(\dfrac{\partial H}{\partial P}\right)_T = -T\left(\dfrac{\partial V}{\partial T}\right)_P + V = -RT/P + RT/P = 0$

7.5 $\left(\dfrac{\partial H}{\partial P}\right)_T = -T\left(\dfrac{\partial V}{\partial T}\right)_P + V = -T\left[\dfrac{\partial[RT/p - a/T]}{\partial T}\right]$

$= -RT/P + a/T + RT/P - a/T = 0$

7.6 $\quad C_p = C_V \left[\left(\dfrac{\partial E}{\partial V} \right) + P \right] \left(\dfrac{\partial V}{\partial T} \right)_P$

$\qquad = C_V + \left[T \left(\dfrac{\partial P}{\partial T} \right)_V - P + P \right] \left(\dfrac{\partial V}{\partial T} \right)_P$

7.7   a. $\quad \left(\dfrac{\partial E}{\partial A} \right)_T = T \left(\dfrac{\partial S}{\partial A} \right)_T + \gamma = -T \left(\dfrac{\partial \gamma}{\partial T} \right) + \gamma$

     b. $\quad \left(\dfrac{\partial \gamma}{\partial T} \right) = -E_s/T_c = - \left(\dfrac{\partial S}{\partial A} \right)$

$$\Delta S = \int_{A_1}^{A_2} (E_c/T_c) dA = E_c/T_c(A_2 - A_1)$$

7.8   a. $dE - TdS - PdV + P'dD$

     b. $dZ = -SdT + VdP + P'dD$

     c. $\left(\dfrac{\partial Z}{\partial D} \right)_{P'} = P'$

     d. $\left(\dfrac{\partial E}{\partial D} \right)_T = T \left(\dfrac{\partial S}{\partial D} \right)_T + P' = T \left(\dfrac{\partial P'}{\partial T} \right)_D + P'$

7.9 $\left(\dfrac{\partial S}{\partial M} \right) = \phi \qquad \left(\dfrac{\partial S}{\partial N} \right)_T = \mu$

Chapter 9

9.1 $RTd \ln(c_1) - Ld\tau_1 = RTd\ln(c_2) - Ld\tau_2$

     $RT \ln(c_1) - L_1\tau_1 = RT \ln c_2 - L_2\tau_2$

9.2 $-S_1 dT + zFd\psi_1 = -S_2 dT + zFd\psi_2$

     $-H_1 dT/T + zFd\psi_1 = -H_2 \, dT/T + zFd\psi_2$

     $\Delta H \ln (T_2/T_1) = zF(\psi_2 - \psi_1) \qquad \Delta H = H_2 - H_1$

9.3   a. Observed $\pi = \{G/M)RT$

       $0.001 = (.5)/M \; 25 = 12.5/M$

       $M = 12.5/0.001 = 12500$

     b. $12500 = 50,000/\text{total ions} \qquad$ Total ions $= 4$ (1 protein, 3 K^+)

9.4 $[Na]_2[Cl]_2 = [Na]_1[Cl]_1$

$(0.11)(0.11) = (0.1 + x - 0.01)(0.1 - 0.01) = 0.09(0.09 + x)$

$0.134 = 0.09 + x \qquad x = 0.044$

9.5 $\Delta G = 0 = V_m(1-P) + 75 + V_r(P-1)$ 3 step cycle

$75 = (V_m - V_r)(P-1) = 16.3 - 15.5) \times 10^{-3} \ (P-1)$

$P-1 = 93.8 \times 10^3$

9.6 $V_l dP_l = V_g dP_g = RT \ dP/P$

$2.46(11-1) = 0.082 \ (300) \ \ln(P/1)$

$1 = \ln(P/1) \qquad P = 2.7 \text{ atm}$

9.7 $V\pi = -RT \ \ln(X_2/X_1)$

$P-(P + .5) = -0.5 = -2500 \ V^{-1} \ 2500 \ \ln(X_2/X_1)$

9.8 $c_2 = [K\}_2 = [Cl]_2 \quad [Cl_1] = c_1 \quad [K] = c_1 + zc_p$

$C_2{}^2 = c_1[c_1 + zc_p]$

9.9 a. $RT \ \ln(X) + \gamma dA = RT \ \ln(1) = 0$

b. $120 \ (1.1-1) = -8.3 \times 300 \ \ln(X/1) \quad \ln(X) = -0.048$

$X = 0.953$

9.10 $$RTd \ \ln(c_L) + zFd\psi_L = RTd \ \ln(c_H) + zFd\psi_H$$

$$- RT \ \ln(c_H/c_L) = zf(\psi_H - \psi_L)$$

Decrease $RT \ \text{ln} c$ must still equal $m'gh$ so c must be smaller

Chapter 10

10.1 $5!/2!3! / 2^5 = 10/32 = 5/16$

10.2 $$d/dn_i[\ \ln(N!/n_1!n_2!n_3!) - + \alpha(N - n_1 - n_2 - n_3)] = 0$$

For 1, 2 or c

$Ln(n_i) - \alpha = 0 \quad n_i = \exp(\alpha) \quad N = 3\exp(\alpha) \quad$ each $n_i = N/3 = \exp(\alpha)$

Chapter 11

11.1 a. $q = 1 + 0.2 + 0.04 = 1.24$

b. $p = 0.24/1.24 = 0.19$

        $<\text{quanta}> = [0 \times 1 + 1 \times 0.2 + 2(.04)]/1.24 = 0.28/1.24 = 0.23$

11.2 a. $[1/1.6](80) = 50 \qquad [0.4/1.6](80) = 20 \qquad [0.2/1.6](80) = 10$

b. $<\varepsilon> = 5/8(0) + 2/8(1kT) + 1/8(2kT) = 0.5kT$

c. $\Omega_{max} = 80!/50!20!10!$

11.3 a. $1/(1+0.6+0.4) = \frac{1}{2}$

b. $<\varepsilon> = 0.6/1(0.5kT) + 0.4/1(0.9kT) = 0.66kT$

$<A> = -kT\ln(2)$

$<s> = [<E> + <A>]/T = 0.66\ \mathrm{k} + kT\ln(2) = 1.35\ \mathrm{k}$

11.4 a. $p_0 = \dfrac{\exp(-\beta\varepsilon_o)}{\exp(-\beta\varepsilon_o) + \exp(-\beta\varepsilon_1) + \exp(-\beta\varepsilon_2)} = \dfrac{\exp(-\beta\varepsilon_o)}{q}$

b. $<\varepsilon> = \dfrac{\exp(-\beta\varepsilon_o)\varepsilon_o + \exp(-\beta\varepsilon_1)\varepsilon_1 + \exp(-\beta\varepsilon_2)\varepsilon_2}{q}$

c. $<A> = -kT\ln(q)$

d. $\dfrac{q^{10}}{10!}$

Chapter 12

12.1 a. $\xi = 1 + 2\lambda q_b + \lambda^2 q_b^2$

b. $<\text{none bound}> = 1/\xi$

12.2 $q = 1 + x + x^2 + \ldots = 1/(1-x) \qquad x = \exp(-\beta E)$

$<\varepsilon> = [0*1 + Ex + 2Ex^2\ldots]/q$

$<A> = -100kT\ln(q)$

12.3 a. Empty sites; 3 SSSA   3 SSASA   1 SASASA

b. $\Xi = (1+\Lambda q_b)^3 \qquad \lambda = \exp(\beta\mu) \qquad q_b = \exp(\beta\varepsilon_b)$

12.4 a. $\xi = 1 + \lambda + \lambda\lambda'$

b. $p(2\ \text{absorbed}) = \lambda\lambda'/\xi$

c. $<N> = [0(1) + 1(\lambda) + 2(\lambda\lambda')]/\xi$

d. $\Xi = \xi^3$

12.5 a. $G_1G_2 \qquad C_1C_2 \qquad G_1(G_2C_1)C_2 \quad C_1(C_2G_1)G_2 \qquad (C_1G_1)(C_2G_2)$

b. $\Xi = 1 + 2\exp(-\beta E_1) + \exp(-2\beta E_1)$

12.6 a. $p(B) = \exp(-\beta\Delta G)/[1+ \exp(-\bar{\beta}\Delta G)]$

b. $\xi = 1 + \exp(-\beta\Delta G)[1 + 2\lambda q_b + \lambda^2 q_b^2]$

c. $p(B) = \exp(-\beta\Delta G)[1 + 2\lambda q_b + \lambda^2 q_b^2]/\{1 + \exp(-\beta\Delta G)$
$[1 + 2\lambda q_b + \lambda^2 q_b^2]\}$

d. $<Ns> = \exp(-\beta\Delta G)[1\times 2\lambda q_b + 2\times \lambda^2 q_b^2]/[1 + \exp(-\beta\Delta G)$
$[1 + 2\lambda q_b + \lambda^2 q_b^2]$

12.7 $\mu = \left(\dfrac{\partial A}{\partial N}\right) = -\dfrac{\partial[kT\ln Q]}{\partial N} \quad -\dfrac{\mu}{kT} = \left(\dfrac{\partial\ln(Q)}{\partial N}\right)_{M,T}$

$-\dfrac{\mu}{kT} = \left(\dfrac{\partial\ln\left(\dfrac{M!}{N!(m-N)!}q^N\right)}{\partial N}\right) = \ln N - \ln(M-N) + \ln(q) = \ln\dfrac{Nq}{M-N}$

$= \ln\left(\dfrac{\theta q}{1-\theta}\right)$

12.8 a. $f_R = 1/[1 + L\{1 + q_1\lambda_1 + q_2\lambda_2)]$

12.9 a. $R, RS_1, RS_2, S_1RS_2, T$

b. $\xi = 1 + \lambda_1 q_{b1} + \lambda_2 q_{b2} + \lambda_1 q_{b1} \lambda_2 q_{b2} + \exp(-\beta \Delta G)$

c. $f_R = 1 + \lambda_1 q_{b1} + \lambda_2 q_{b2} + \lambda_1 q_{b1} \lambda_2 q_{b2}/\xi$

d. $<N(S_1+S_2) = [1(\lambda_1 q_{b1}) + 1(\lambda_2 q_{b2}) + 2(\lambda_1 q_{b1}\ \lambda_2 q_{b2})]/\xi$

12.10 a. f_s (no magnetic field) $= 1/4$

b. $f_{+\epsilon} = \exp(-\beta \epsilon)/[1+ \exp(-\beta \epsilon_-) + 1 + \exp(-\beta \epsilon_+)]$

Chapter 13

13.1
$$\frac{\int_0^\infty \varepsilon \varepsilon^{1/2} \exp(-\beta \varepsilon) d\varepsilon}{\int_0^\infty \varepsilon^{1/2} \exp(-\beta \varepsilon) d\varepsilon}$$

13.2 a. $$<v^2> = \frac{\int_0^\infty v^2 \exp(\beta m v^2/2)v\,dv}{\int_0^\infty \exp(\beta m v^2/2)v\,dv} \qquad a = \beta m/2$$

b. $<v^2> = \dfrac{1/2a^2}{1/2a} = 1/a = 2/m\beta$

13.3 a. $q = \int_0^\infty \exp(-\beta \varepsilon) d\varepsilon = \dfrac{1}{-\beta} \exp(-\beta \varepsilon)|_0^\infty = 1/\beta$

b. $p = \int_{\varepsilon_a}^\infty \exp(-\beta \varepsilon) d\varepsilon/\dfrac{1}{\beta} = \exp(-\beta \varepsilon_a)$

c. $<\varepsilon^2> = \beta^{-1} \int_0^\infty \varepsilon^2 \exp(-\beta \varepsilon) d\varepsilon$

d. $<\varepsilon^2> = \dfrac{\partial^2 \ln q}{\partial \beta^2}$

13.4 $q = \int_{-\infty}^\infty \exp(-\beta p^2/2m) dp = \sqrt{\dfrac{2m\pi}{\beta}}$

$<\varepsilon> = -\dfrac{\partial \ln q}{\partial \beta} = kT/2$

13.5 a. $(kT)^s = q$

b. $<\varepsilon> = \int_{E_1}^{E_2} \exp(-\beta \varepsilon) \varepsilon^{s-1} d\varepsilon/(kT)^s d\varepsilon$

c. $<\epsilon> = skT$

13.6 $\quad q = \int_0^\infty \exp(-\beta mgh)dh = \dfrac{1}{-\beta mg} \exp(-\beta mgh)|_0^\infty = \dfrac{1}{\beta mg}$

$\quad\quad <h> = (\beta mg) \int_0^\infty h \exp(-\beta mgh)dh$

Chapter 14

14.1 $\quad T = \begin{pmatrix} \lambda_A \exp(-\beta\varepsilon_{AA}) & \lambda_A \\ \lambda_B & \lambda_B \exp(-\beta\varepsilon_{BB}) \end{pmatrix}$

14.2 $\quad I(CaCl_2) = [0.01\,(2)^2 + (.02)(-1^2)]/2 = 0.03$

$\quad\quad K^{-1}(CaCl_2) = 3\ nm(0.01/0.03)^{1/2} = 1.7\ nm$

$\quad\quad I(CaSO_4) = [0.01(2)^2 + 0.01(-2)^2]/2 = 0.04$

$\quad\quad K^{-1}(CaSO_4) = 3\ nm(0.01/0.04)^{1/2} = 1.5\ nm$

14.3 $\quad$ a. $\xi = 1 + 3L + 3L^2 q_i + L^3 q_i^3 \quad\quad q_i = \exp(-\beta\varepsilon_{tt})$

$\quad\quad$ b. $<N_t> = [1(3L) + 2(3L^2) + 3(L^3)q_b^3]/\xi$

$\quad\quad$ c. $<N_{ttt}> = 1(3L^3 q_b^3)/\xi$

Chapter 15

15.1 $\quad dA^*/dt = k_1[A][B] - k_{-1}[A^*][B] - k_f[A^*] = 0$

$\quad\quad$ a. $[A^*] = k_1[A][B]/[k_{-1}[B]] + k_f$

$\quad\quad$ b. $I = k_f[A^*]$

15.2 $\quad$ a. $A = 3t, 2r, 1v \quad\quad A^* = 3t, 2r, 1 - 1 = 0v$

$\quad\quad K^{\ddagger} = q_t^3 q_r^2/q_t^3 q_r^2 q_v = q_v^{-1} = 0.1$

$\quad\quad k_2 = 10^{13} \times 0.1 = 10^{12}$

$\quad\quad$ b. $<A> = -RT\ln(K) = -2500\ln(0.1)$

15.3 $\quad K = \dfrac{q_t^3 q_r^3 q_v^2}{q_{qt}^3 q_r^2 q_v^1 q_t^3}$

15.4 $\quad$ a. Activated complex has two translations, 1 rotation and 1 vibration

$\quad\quad$ b. $K^{\ddagger} = \dfrac{q_t^2 q_r^1 q_v^0}{[q_t]^4} \quad\quad k_2 = K^{\ddagger}(kT/h)$

15.5   a. $\dfrac{(N+j+k-1)!}{N!(j+k-1)!} = D$ all states for complex

b. $\dfrac{(i+g-1)!(N-i+k-1)!}{i!(g-1)!(n-i)!(k-1)!}/D$

15.6   a. $3 \times 10^9\ [A][B] = Z$

b. 50,000 J/mol

Chapter 16

16.1 $J_w = L_{11}\ dP/dx + L_{12}\ d\pi/dx$

$J_s = L_{21}\ dP/dx + L_{22}\ d\pi/dx$

Moving ions carry some water, moving water carries some ions

16.2 $J_{heat} = L_{11}\ dT/dx + L_{12}\ d\psi/dx$

$J_{charge} = L_{21}\ dT/dx + L_{22}\ d\psi/dx$

16.3 $<x^2> = [(-4^2)1 + (-2)^2 4 + 0^2(6) +(+2)^2(4) +(4^2)1]/16 = 64/16 = 4$

16.4 $J_T = (1/A)dS/dt \qquad X_t = dT/dx$

$J_P = (!/A)dV/dt \qquad X_P = dP/dx$

Index